Liberation Technology

A *Journal of Democracy* Book

•

Published under the auspices of
the International Forum for Democratic Studies

Liberation Technology
Social Media and the Struggle for Democracy

Edited by Larry Diamond and Marc F. Plattner

The Johns Hopkins University Press
Baltimore

9 8 7 6 5 4 3 2 1

Chapters in this volume appeared in the following issues of the *Journal of Democracy:* chapter 1, July 2010; chapter 2, October 2010; chapters 4, 5, and 6, April 2011; chapter 8, July 2011. For all reproduction rights, please contact the Johns Hopkins University Press.

The Johns Hopkins University Press
2715 North Charles Street
Baltimore, Maryland 21218-4363
www.press.jhu.edu

Library of Congress Cataloging-in-Publication Data

Liberation technology : social media and the struggle for democracy / edited by Larry Diamond and Marc F. Plattner.
    p. cm. — (A journal of democracy book)
    Includes bibliographical references and index.
    ISBN 978-1-4214-0567-4 (hdbk. : alk. paper) — ISBN 1-4214-0567-9 (hbk. : alk. paper) — ISBN 978-1-4214-0568-1 (pbk. : alk. paper) — ISBN 1-4214-0568-7 (pbk. : alk. paper) — ISBN 978-1-4214-0698-5 (electronic) — ISBN 1-4214-0698-5 (electronic)
    1. Political participation—Technological innovations. 2. Democratization—Technological innovations. 3. Social media—Political aspects. I. Diamond, Larry Jay. II. Plattner, Marc F., 1945–

    JF799.L53 2012
    303.48'33—dc23
                                 2012012206

CONTENTS

IV. Policy Recommendations

ACKNOWLEDGMENTS

In the most immediate sense, this book originated in a conference on "Liberation Technology in Authoritarian Regimes" held at Stanford University in October 2010. Eight of our eleven chapters were initially presented at that conference, and the first chapter in this volume helped to frame the conference. We are indebted to the Center on Democracy, Development, and the Rule of Law (CDDRL) at Stanford for organizing and hosting that conference, to Steven Kahng for his financial support of the meeting, and to the Lynde and Harry Bradley Foundation for its generous support of the Liberation Technology Program at CDDRL. We would also like to thank Larry Diamond's faculty colleagues in that program, Joshua Cohen and Terry Winograd, for their role in helping to shape the research agenda and the conference, and Anna Davies for her research and programmatic assistance in launching the program, as well as in helping to organize the conference itself.

In a deeper sense, however, we were pushed, prodded, and inspired into producing this book by our students and our younger colleagues and fellows, who became deeply immersed in social media well before we, as editors and scholars, appreciated the political and social significance and the liberating potential of these burgeoning technologies. They are too numerous to name, but we would like to thank them all. In particular, we thank the following Stanford students: Taylor Dewey, for his critical editorial assistance and organizational and research support in the preparation of this manuscript; Astasia Myers, for her superb research and editorial assistance; Blake Miller, for his eye-opening field research and subsequent assistance on China; and Lucinda Gibbs, who also provided energetic research assistance and support for the Liberation Technology Program.

As with the two dozen earlier books in this series (particularly those that have included previously unpublished chapters along with articles that had already appeared in the *Journal of Democracy*), the *Journal* staff has had to juggle the editing and production of this new material with its work on quarterly issues of the *Journal*. For his patient and artful management of the production and layout, we once again thank most heartily Managing Editor Brent Kallmer. Assistant Editor Marta Kalabinski conducted the final editing of two of the previously unpublished chapters with intelligence and skill, under pressure of time. Executive Editor Phil Costopoulos and Associate Editor Tracy Brown once again left a deep imprint on our book, not only by editing previously unpublished manuscripts, but also by helping to make the *Journal* articles published here considerably more lucid and accessible than they might otherwise have been. Together, these four individuals constitute a great editorial team, and we have been privileged to work with them for several years. We also once again thank Dorothy Warner for her care and precision in compiling the index.

This is the last book we will produce as part of our two-decades-long collaboration with Henry Y.K. Tom, who directed social science book publishing at the Johns Hopkins University Press (JHUP), and in that capacity recruited and steered to publication many of the most influential works in comparative democratic studies. We discussed the idea for this book with Henry, and he was enthusiastic about its prospects, but he passed away suddenly in January 2011, well before it took final shape. We are very indebted to his successor as editor of JHUP's books on democratization, Suzanne Flinchbaugh, who immediately recognized the urgent importance of the topic and encouraged us to bring the project to a timely completion. We also thank our colleagues at JHUP's Journals Division, Bill Breichner and Carol Hamblen, for our continuing valued partnership with them in the production of the *Journal of Democracy*.

There are many others whom we are pleased to thank for their contribution to our efforts, beginning with NED's president Carl Gershman and its Board of Directors (chaired by Richard A. Gephardt), who have given us steady support along with the utmost respect for the *Journal*'s editorial independence and integrity. We also want to express our thanks to the many members of the NED staff who have lent the *Journal* their expertise in editorial matters large and small, and to NED's Center for International Media Assistance, which has helped us to understand the contribution of the new digital media to democratization. And we remain especially grateful for the financial support that the *Journal* has received since its inception from the Lynde and Harry Bradley Foundation.

INTRODUCTION

Larry Diamond

Few, if any, developments in recent decades have more profoundly transformed politics and civil society than the emergence of digital information and communication technologies (ICTs). Prominent among these have been the Internet; the sprawling blogosphere that it has spawned; the proliferating array of social-media tools such as Facebook, Twitter, YouTube, and Flickr; and the galloping growth in access to these digital media through mobile phones. These electronic tools have provided new, breathtakingly dynamic, and radically decentralized means for people and organizations to communicate and cooperate with one another for political and civic ends.

The surge in use of these digital ICTs has been dizzying. In fact, most of them have become pervasive only in the last few years. The number of Facebook users in the world exploded from 12 million in December 2006 to 100 million in August 2008, to half a billion in July 2010. By early 2012, it was well in excess of 800 million.[1] By the time you are reading this book, more than a billion people will likely be using Facebook, with most of the recent growth coming in emerging-market countries. The number of tweets (Twitter messages) per day has increased from 300,000 in early 2008 to 2.5 million in 2009, 35 million in 2010, and 200 million by mid-2011.[2] YouTube went from 8 million views of its videos per day when it launched in December 2005 to 100 million per day in July 2006, to a billion views per day in October 2009—and twice that by May 2010, just seven months later.[3] A year after that, YouTube's daily views were at three billion. As Peter Diamandis, chairman of the X Prize Foundation and author of the new book *Abundance* notes, a Masai warrior standing on the Kenyan plains with an ordinary mobile phone has better access to mobile communications than U.S. president Ronald Reagan had in the mid-1980s, and if he has a smart phone he has wider access to information online than President Bill Clinton had in the mid-1990s.[4]

Most of those who use ICTs for political and civic purposes are not political activists. Rather, they consume and exchange information and opinions online. But in the digital age, the lines between reader and reporter, news and opinion, and information and action have all become blurred. Between late 2006 and the end of 2011, the number of blogs in the world increased fivefold, from 35 million to 181 million.[5] Most bloggers and tweeters focus not on politics but on family, friends, culture, consumption, and other personal and parochial matters. In the last six years, however, social media have become essential to election campaigns in developed and emerging democracies alike. In the United States, Barack Obama may well owe his presidency to his campaign's use of the Internet, social media, and mobile-phone messaging to raise money and mobilize and connect grassroots volunteers in ways more effective than any of his rivals—or any candidate in history—could manage.

Digital communication technologies are transforming elections, political debate, civic advocacy, philanthropy, and the structure of the mass media in contemporary democracies. Today, the arena for political commentary and competition is more fast-paced, more decentralized, and more open to new voices and social entrepreneurs than ever before. Yet many observers worry that it is also more fragmented and polarized, as people sort themselves into ideologically exclusive milieus in which they are seldom if ever exposed to sharply divergent points of view. Thus, there is an ongoing analytic debate about whether the digital age is improving or diminishing the quality of democratic politics. Probably it is doing some of both.

A parallel debate is occurring over the impact of digital ICTs on authoritarian regimes. Do the Internet, social media, mobile phones, and their exploding array of applications empower citizens to mobilize for freedom and accountability, or do these technologies empower autocracies to better monitor and effectively neutralize prodemocracy movements and dissidents? What are the roles and responsibilities of private companies that sell Internet surveillance and control technology to authoritarian states, and also increasingly monitor and shape the online activity of individual users? Do digital tools distract citizens from political action by promoting personal entertainment, or otherwise weaken democratic resistance by giving people a seemingly safe but superficial illusion of political action through digital means? And if digital ICTs do indeed represent potent new tools for both democrats and autocrats, which force—regime or opposition—is gaining the upper hand?

It is to this second set of questions that this volume is addressed. Several years ago, as I was completing a work on the worldwide struggle for democracy, I became struck by the growing use of the Internet, the blogosphere, social media, and mobile phones to expose and challenge the abuses of authoritarian regimes; to provide alternative channels through which information and communication could flow outside the censorship and controls imposed by dictatorships; to monitor elections; and to mobilize people to protest.

By 2007—which now seems like a generation ago in terms of the speed with which these technologies have developed—digital ICTs had already registered some stunning successes. The new technologies had enabled Philippine civil society to fill the streets to drive a corrupt president (Joseph Estrada) from power; facilitated the rapid mass mobilizations against authoritarianism mounted by the Orange Revolution in Ukraine and the Cedar Revolution in Lebanon, respectively; documented the rigging of the 2007 elections in Nigeria; exposed (via satellite photography) the staggering inequality embodied in the vast palace complexes of Bahrain's royal family; and forced the suspension of an environmentally threatening chemical plant in Xiamen, China, through the viral spread of hundreds of thousands of impassioned mobile-phone text messages. I called the ICTs that these citizens were using "liberation technology" because of their demonstrated potential to empower citizens to confront, contain, and hold accountable authoritarian regimes—and even to liberate societies from autocracy.[6]

The term "liberation technology" has since become widely used and hotly contested. Critics assert that it contains an analytic bias in presuming that digital ICTs are intrinsically "liberating"—that is, inherently good for promoting freedom, accountability, and democracy. Yet, as I argue in the next chapter, we can apply the term to any form of ICT "that can expand political, social, and economic freedom," while remaining agnostic and evidence-driven in seeking to determine whether and under what conditions it *actually* does so. In this volume—which had its origins in a Stanford University conference on "Liberation Technology in Authoritarian Regimes"—we maintain this empirical and open-minded approach. Some of the contributors to this volume—particularly Xiao Qiang, Patrick Meier, Philip Howard and Muzammil Hussain, Walid Al-Saqaf, and Mehdi Yahyanejad and Elham Gheytanchi—view the question from the standpoint of the users themselves. On the whole, these authors take heart from the tenacity and innovative spirit that individuals and civil society groups in China and the Middle East have shown in using liberation technology to find ways around authoritarian constraints. Others among the present volume's authors—Ronald Deibert, Rafal Rohozinski, Evgeny Morozov, Rebecca MacKinnon, and in some respects Walid Al-Saqaf—take authoritarian regimes as their starting point, and express concern at how fast such regimes are moving as they innovate, build capacity, and harden resolve to control digital tools.

Yet all the authors share a common normative perspective—that the Internet should be a universal, open, and essentially free space; and that people should be able to use digital ICTs to report, document, advocate, exchange, and organize peacefully for political change, while having their rights (including the right to privacy) respected. This means not that regulation should be absent, but rather that it should be carefully constrained—used to protect intellectual property rights and shield vulnerable groups such as children, for example. The intersection of this normative perspective and the many empirical developments documented in this book generates a diffuse agenda for international policy, which is the subject of Daniel Calingaert's concluding chapter.

Liberation vs. Control in Cyberspace

My opening chapter sets the scene for the later chapters by introducing the phenomenon of liberation technology. Technology is liberating, I argue, to the extent that it can "empower individuals, facilitate independent communication and mobilization, and strengthen an emergent civil society." Empowerment is not only political. To the extent that ICTs better enable individuals to manage their health, educate their children, locate resources, be secure in their persons, or sell their produce at competitive rates, ICTs may also be profoundly empowering. In some of the world's poorest places, liberation technologies, using the mobile phone as a platform, are helping to lift people out of poverty by giving them valuable information and improving the openness and efficiency of markets. This engages the vast, burgeoning field of "ICT4D"—where the "4D" means "for development."[7] In this volume, we are limiting our focus to the impact of ICTs on the "other D"—democracy.

Over the past decade, we have seen a vast array of instances in which people have used ICTs and the new social media to report, expose, organize, and protest outside of the normal authoritarian constraints. The chapters that follow are replete with examples from China and the Middle East, but I also note the important role that alternative online journalism is playing in countries such as Malaysia to undermine the authoritarian regime's monopoly of information and to expose corruption, human-rights abuses, ethnic discrimination, and police brutality. In Africa and throughout the developing world, the mobile phone is transforming politics by enabling more efficient and comprehensive election monitoring; improving budget transparency and accountability; informing citizens about efforts to combat corruption and human-rights abuses; and mapping such abuses as well as outbreaks of interethnic violence. A crucial facilitator of these efforts has been the rapid development and diffusion of open-source software such as FrontlineSMS (which enables large-scale, two-way text messaging) as well as the Ushahidi "crisis-mapping" platform described by Patrick Meier in Chapter 7.

I stress, as in one way or another do most of this volume's authors, that "technology is merely a tool, open to both noble and nefarious purposes." This book does not much address (beyond Chapter 2) the dark sides of ICT usage—for online crime, child pornography, terrorism, cyberwarfare, and the theft of corporate and government secrets. We do deal extensively with the ways in which governments are using these technologies, or turning them inside out, to monitor online opposition, suppress dissent, censor and remove "harmful content," and identify and arrest leading critics. Indeed, one can find a disturbing pattern in the use of liberation technology in highly authoritarian contexts such as Burma in 2007, Iran in 2009, and Bahrain in 2011. Citizens use mobile phones and social media to mobilize protests and to film them, thereby expanding domestic and international support. Then the regime cracks down brutally and protests shrivel, while state security services mine online photographs and films, heighten Internet filtering and surveillance, and even use ICTs proactively to track down and arrest protest leaders.

The balance of potency between ICTs as democracy-boosters and ICTs as repression-enablers remain dynamic and fluid. One key factor is the ability of opposition groups to acquire, master, and improve censorship-evading tools such as those that Walid Al-Saqaf discusses in Chapter 9. I remain, on balance, optimistic about liberation technologies' potential to raise democratic consciousness and capacities and ultimately to promote democratic transitions in authoritarian regimes. At the same time, however, I am also keenly aware that "there is now a technological race underway between democrats seeking to circumvent Internet censorship and dictatorships that want to extend and refine it." As Daniel Calingaert explains at length in Chapter 11, much will depend on what international actors (especially democratic governments) do or fail to do to ensure free and unfettered access to the Internet and to restrict authoritarian access to advanced filtering and surveillance technologies.

In Chapter 2, Ronald Deibert and Rafal Rohozinski illuminate the dark side of liberation technologies and challenge the dominant liberal-democratic vision of them by showing how they are increasingly used for malicious purposes. For example, al-Qaeda remains viable largely because of *jihadi* websites—"jihadists and militants mobilize around a common 'imagined community' that is nurtured online." It is not that cyberspace is an "ungoverned realm," but rather that it is a weakly governed space filled by a complex and rapidly evolving mix of actors with widely divergent motives. Their analogy is "to a gangster-dominated version of New York: a tangled web of rival public and private authorities, civic associations, criminal networks, and underground economies." In this cyberstew, liberation and control, transparency and deception, cooperation and predation, tolerance and extremism all vie with one another, just as they do in physical space. But we should not assume that benign and liberating purposes will trump repression and exploitation in cyberspace. Deibert and Rohozinski report that the production of "malware" (malicious software that can gain control of an unsuspecting user's computer for the purposes of crime, surveillance, or sabotage) "is now estimated to exceed that of legitimate software," and the growth of malware is increasing at an alarming rate.

Authoritarian governments are not only monitoring, filtering, and censoring the Internet, they may also have launched or encouraged the coordinated attacks that human-rights and prodemocracy groups have experienced in recent years in China, Burma, Tibet, and numerous countries of the former Soviet Union. Moreover, as democratic governments themselves regulate the Internet and establish cyberwarfare capacities, authoritarian governments find it easier to justify their more radical measures. And they have at their disposal a growing variety of "next-generation" controls that go well beyond Internet filtering. These include legal measures to restrict, intimidate, and prosecute independent reporting and critical commentary online; pressuring private Internet service providers (ISPs) to do the dirty work of monitoring online activity, removing challenging content, and blocking troublesome organizations; selective, "just-in-time" blocking, including through distributed denial-of-service (DDoS) attacks designed to cripple opposition organizations at key political moments; as well as "patriotic hacking" (by Chinese and Iranian cybermilitants, for example), targeted surveillance, and social-malware attacks. Deibert and Rohozinski conclude that the quest for liberal governance of cyberspace, where freedom and individual privacy and security are effectively protected, "will not come overnight with the invention of some new technology. Instead, it will require a slow process of awareness-raising, the channeling of ingenuity into productive avenues, and the implementation of liberal-democratic restraints."

In Chapter 3, Ronald Deibert widens the optic of repression to examine the international mechanisms and dynamics of cyberspace controls. While most efforts to monitor controls in cyberspace (including the OpenNet Initiative [ONI], for which Deibert is a principal investigator[8]) have focused on the policies and actions of individual states, Deibert emphasizes that states compete with one another, and that they also "borrow and share best practices, skills, and technologies." The technical tools of Internet filtration appear to be diffusing across borders, and numerous states appear to have taken a cue from one another by disabling SMS (short-message service, or mobile-phone text messaging) and instant-messaging services during periods of intense opposition challenge or protest. By the same token, civil society organizations increasingly cooperate and share technologies and best practices across borders, and many civil society networks seeking to promote human rights, freedom, and democracy are truly international. Then there are the private-sector actors—not least the US$80-billion-per-year cybersecurity market—who seek commercial opportunities and profit throughout the world. The international trend, Deibert finds with concern, is "a growing norm worldwide for national Internet filtering." And as noted in the previous chapter, "the private-sector actors who own and operate the vast majority of cyberspace infrastructure are being compelled or coerced to implement controls on behalf of states." Part of the work of the OpenNet Initiative is to document the collusion of Internet companies in restricting access in authoritarian states and to bring "pressure on those companies to become more accountable." The same goes for companies that are eagerly producing the hardware and software for an "exploding market" in Internet surveillance, control, and sabotage, including "deep packet inspection and traffic shaping capabilities" that violate principles of network neutrality.

Deibert urges new analytic attention to the international institutions for Internet governance: the International Corporation for Assigned Names and Numbers (ICANN), the International Telecommunication Union (ITU), and the Internet Governance Forum (IGF). Increasingly, governments that want to restrict Internet access are taking these forums "seriously as vectors of policy formation and propagation," and scholars must to do the same. Countries such as China and Russia now engage in these various international forums in order "to reassert the legitimacy of national sovereign control over cyberspace by promoting such a norm at international venues." As a result, deliberations in these bodies are increasingly being politicized and infused with national-security considerations, an ominous new trend. Deibert also shows how regional bodies like the Shanghai Cooperation Organization are helping to coordinate and advance these authoritarian national strategies of Internet surveillance and control, and he observes a growing pattern of bilateral cooperation, with China serving as a major source of export or diffusion of filtering and surveillance technologies.

Governments also legitimize far-reaching Internet-control policies under the guise of merely doing what democracies like the United States do to monitor and combat terrorism or protect copyrights. No less worrisome is the "arms-race spiral in cyberspace," as dozens of governments build up cyberwarfare capabilities in response to what they see other states, particularly China and the United States, to be doing. Deibert concludes with a warning: "General statements about the 'war on terror' or 'copyright controls' can be turned into excuses for a broad spectrum of otherwise nefarious actions by authoritarian regimes," and therefore democratic regimes must set an international example of consistency, transparency, accountability, and fidelity to their own liberal values.

Deibert's concerns are shared by Evgeny Morozov, who warns in Chapter 4 of a number of ominous trends that not only threaten Internet freedom but utilize new breakthroughs in digital technology to imperil freedom more generally. Technological forms of Internet control are not only growing in power and subtlety, but they are being supplemented by sociopolitical forms of control that can have an even more chilling effect—for example, by holding blogging platforms legally responsible for all the content on their sites, or arresting and even torturing those who exercise their personal freedom on the web. Framing the challenge as simply a technological arms race in which more resources for additional technical innovations will advantage advocates of freedoms can be a dangerous delusion, Morozov cautions. Technical fixes, including censorship-circumvention software like Tor, do nothing to restrain autocracies from other methods of repressing freedom online, many of which are much harder to document and trace than Internet filtration. One example is the rising incidence of DDoS attacks, which "target individuals or entire organizations by flooding their websites with crippling volumes of artificially generated Internet traffic." These not only shut down the targeted site for all users for some period of time, they also impose heavy psychological, financial, and operational costs on the organization attacked, while leaving few if any clear footprints as to responsibility. They may thus be even more effective than censorship.

Wily and resourceful authoritarian regimes, like Vladimir Putin's in Russia, are also perfecting the dark and subtle arts of socially infiltrating online networks with saboteurs "to create artificial splits within the community" and provoke "community administrators to take harsh and unpopular measures." States are also asserting "informational sovereignty," in essence moving to detach their national cyberspace from the global Internet by constructing their own national e-mail systems, search engines, and versions of social media like Twitter and Facebook, as China has already done to a considerable extent.

Like Deibert, Morozov worries about the trend toward commercial outsourcing of Internet control to private companies that are forced to take on a broad "self-policing" role as the price of doing business in the country. This may even prove more effective than direct state control, he observes, because the commercial Internet companies "are more likely to catch unruly content, as they are more decentralized and know their own online communities better than do the state's censors." Moreover, in the intense competition for lucrative profits, private companies are rapidly innovating in the development of Orwellian tools of surveillance, such as powerful new software for indexing and searching video-surveillance footage, and face-recognition software that can identify individuals and then search social-networking or other websites for photos of the same person.

As search-engine companies such as Google and a wide variety of other Internet firms develop the capacity to permanently store and mine vast amounts of data on the Internet behavior—and thus beliefs and preferences—of tens or hundreds of millions of individuals, the possibilities grow for game-changing leaps forward in the depth and sophistication of Internet censorship and control. These new forms "marry artificial intelligence and basic social-networking analysis" in order to fine-tune control in ways that do not broadly impede economic competitiveness, as the cruder instruments of censorship do. Thus Morozov, like Deibert, urges the governments of the United States and other democracies to be mindful of how their own legal efforts to control crime, piracy, and warfare on the Internet might create "an enabling environment for authoritarian governments that are keen on passing similar measures, mostly for the purpose of curbing political freedom."

Liberation Technology in China

If any one country is going to weigh decisively in settling the debate about whether digital ICTs are more liberating or controlling, it will be China. Some of the reasons are obvious. China is by far the most populous authoritarian country in the world. It is also widely seen as a model of rapid development (and political stability) without democracy. The Chinese Communist Party (CCP) claims to have found a better way, outside the Western democratic model, of effective and responsive governance without the conflict, messy delays, and periodic political deadlock that often go with democracy. Yet as the next two chapters reveal, political protest is rising in China—not least on the Internet. And Chinese who are seeking to defend their rights, challenge corruption and collusion, and extend accountability are making intensive use of the Internet, SMS, and various social media.

Moreover, China now has the largest number of Internet users in the world, over half a billion. That is roughly twice as many as in the United States, more than four times as many as in India, and nearly a quarter of all Internet users in the world. While the proportion of the Chinese population using the Internet (a little less than 40 percent) is still low by the standards of developed economies (where Internet penetration tends to be around 75 to 80 percent), it is now sizeable and growing fast (by more than 50 million users in 2011 alone). And a large number of these users—an estimated 350 million—are able to access the Internet with their mobile phones.[9] By one estimate, China has nearly 200 million "microbloggers" (who post short blog posts, rather like Twitter updates, except that Twitter itself is banned in China).[10]

In Chapter 5, Xiao Qiang, founder and chief editor of *China Digital Times,* examines how Chinese Internet users—"netizens"—are using digital tools to extend the boundaries of expression and generate new, autonomous forms of political participation and dissent, thereby "changing the rules of the game between state and society." While monopoly control of information has been a major pillar of CCP rule since 1949, this has altered with the rapid emergence of networked digital communications that make total censorship and control impossible.

For Xiao, the 2007 grassroots environmental protests against a proposed chemical factory in the coastal city of Xiamen, Fujian Province—organized and spread through e-mail, instant messages, SMS, and cellphone photos—marked the rise of a new phenomenon in China: popular opinion. For the first time, but certainly not the last, ordinary Chinese citizens were able to help shape the policy agenda by using digital media (with an important assist from sympathetic elements in the traditional media). "The rise of blogging, instant messaging, social-networking services," search engines, RSS (web feeds of syndicated content), and Internet bulletin-board systems "have given netizens an unprecedented capacity for communication" in China, he writes. This has enabled the rise of an autonomous "quasi-public space," where "millions of users can generate, distribute, and consume content," and where social and political issues can be discussed "in far bolder language than would be permitted in the official media." And it has inspired the kinds of local protests that halted construction of the chemical plant in Xiamen.

Xiao does not downplay the extraordinary scope and vigor of online censorship—"a complex web of regulations, surveillance, imprisonment, propaganda, and the blockade of hundreds of thousands of international websites" that has come to be known as the "Great Firewall of China." Indeed, it can black out a story when it acts totally and preemptively (as with Liu Xiaobo's award of the 2010 Nobel Peace Prize). But Xiao documents a variety of ways in which Chinese netizens evade state censors, reveal sensitive information about officials, play on words and symbols, employ metaphors and coded language, and employ various forms of popular culture (satire, jokes, songs, and so on) to criticize and even ridicule the ruling party. These can cascade uncontrollably—"like water gushing through a hole in the dam." Even if some of the estimated fifty-thousand Internet police purge them from one place, they crop up in another, quickly becoming "public knowledge"—to the point where the government must sometimes acknowledge and redress a scandal. Thus, he believes that digital ICTs will in the long run liberate China. "Already," he observes, "we are starting to see compromise, negotiation, and rule-changing behavior" in response to grassroots digital mobilization. As grassroots online voices converge with "liberal elements within the established media . . . they are creating a substantial force that is slowly wearing away at the CCP's ideological and social control."

Rebecca MacKinnon shares Xiao Qiang's hopes, but not his optimistic assessment. In Chapter 6, she sees in contemporary China a new model of "networked authoritarianism," in which, for some time to come, digital ICTs are more likely to sustain than erode the hegemony of the CCP (and possibly of other repressive regimes that can copy its techniques). The key innovation of networked authoritarianism is the coexistence of extensive digital communications, including limited space on websites and social-networking services to criticize government policies and social problems, with systematic and technically sophisticated state surveillance and control. The CCP, she argues, has "adapted to the Internet much more successfully than most Western observers realize." The Chinese regime has a relationship with the Internet like that of a human being's with water, a substance that is both "vital and dangerous." Learning as they go, China's Internet police go to great and creative lengths to minimize the danger of the Web. "All Internet companies operating within Chinese jurisdiction—domestic or foreign—are held liable for everything appearing on their search engines, blogging platforms, and social-networking services." This provides a strong incentive for companies to police and censor their own Internet users around the clock.

China deploys a wide range of other repressive tactics: cyberattacks; device and network controls to track and block Internet activity at more localized points; domain-name controls to expunge unauthorized and, especially, anonymous websites; localized disconnection and restriction of the Internet during moments of protest; relentless surveillance of Internet and mobile-device users; and "proactive efforts to steer online conversations." All of these methods serve a common purpose: to preempt, purge, and punish politically threatening digital activity.

At the same time, China's leaders are of course also using digital tools energetically to push their own propaganda. Under this new model of authoritarianism, "the average person with Internet or mobile access has a much greater sense of freedom," and abundant digital fun, but "even most of China's best and brightest are not aware of the extent to which their understanding of their own country—let alone the broader world—is being blinkered and manipulated." And when individuals do press on matters of real political significance, rights of expression and organization are still severely constrained. "Those whom the rulers see as threats are jailed; truly competitive, free, and fair elections are not held; and the courts and legal system are tools of the ruling party." China's emergent civil society may be winning some local battles for accountability and justice, but the authoritarian party-state is winning its war for political survival.

MacKinnon is not only worried about the trends in China. She sees networked authoritarianism as a model of rule that can be replicated, and whose tools and strategies (particularly the legal, regulatory, and political tools beyond Internet filtration) are in fact being replicated by Russia, other post-Soviet republics, and some Middle Eastern regimes, as the authors of previous chapters have noted. Like Deibert, Rohozinski, and Morozov, MacKinnon stresses the importance of democratic governments setting a good example by avoiding pernicious practices like intermediary liability and warrantless surveillance, and by adopting Internet and telecommunications policies that are "transparent, accountable, and open to reform" by both the courts and the political process. Without these international standards, democratic movements "will face an increasingly uphill battle against progressively more innovative forms of censorship and surveillance."

To some extent, the debate between Xiao and MacKinnon is a classic instance of whether one sees the glass as half full or half empty. Neither expects the imminent demise of authoritarian rule. Each sees in China today a mixed picture, in which a rapidly growing body of "netizens" enjoys some autonomy to generate and share information. But while Xiao believes that digital ICTs are truly empowering citizens and eroding the foundations of Communist Party control, MacKinnon sees a new era—and indeed a new model—of much more sophisticated, adaptive, and thus resilient authoritarian rule.

Liberation Technology in the Middle East

If China's Communist rulers remain obsessed with the threat that digital media pose to their rule, it is not without good reason. During 2011, youth-based opposition movements used these digital tools to mobilize popular outrage and protests that quickly brought down two of the oldest and seemingly most stable autocracies in the Arab world, in Tunisia and Egypt. So shaken were the Chinese authorities by these rapidly unfolding events that (as Xiao Qiang reports) they blocked the word "Egypt" from a major Chinese search engine and then, when netizens began to call online for prodemocracy demonstrations like those in Tunisia's Jasmine Revolution, authorities blocked the word "Jasmine" as well.

It is hard to tell how much of a difference liberation technology made to the popular movements that in 2011 toppled long-ruling autocrats in Tunisia and Egypt and stirred widespread protest and revolt in Bahrain, Yemen, Libya, and Syria; or to the movement that had nearly toppled Iran's Islamist authoritarian regime two years earlier after fraudulent elections prompted street protests on an unprecedented scale. Each chapter in this section of our book, however, shows that digital tools played an important, even crucial role, in enabling previously weak opposition forces to rapidly mobilize challenges to authoritarian rule.

In Chapter 7, Patrick Meier focuses on a specific digital tool, crisis mapping, and the role that it played in Egypt's November–December 2010 parliamentary elections. This polling, which the regime shamelessly manipulated to pad its already outsized legislative majority, formed the immediate prelude to the popular revolt against President Hosni Mubarak that broke out on 25 January 2011. Meier, a social scientist who also serves as Director of Crisis Mapping for the web-based mapping platform Ushahidi, seeks to determine whether Ushahidi made a difference in deterring electoral fraud. A digital tool that has been used to map human-rights violations, electoral malpractices, natural-disaster damage, corruption instances, and many other dimensions of crisis and trouble, Ushahidi "is a free and open-source mapping software that allows anyone to create a live and rich multimedia map of an event or situation." Originally developed to map and document Kenya's postelection violence in January 2008, using SMS reports from eyewitnesses, it has since been used to generate more than ten thousand live, event-centered maps in more than forty countries. Having been upgraded over time, it can now be integrated with a wide variety of other digital tools, including Twitter, Facebook, and YouTube, as well as e-mail and voicemail.

Meier analyzes the efforts of a Cairo-based civil society group, the Development and Institutionalization Support Center (DISC), to use what it named "U-Shahid" (*shahid* means "witness" in Arabic) to monitor the parliamentary elections. DISC's specific goals were to help Egyptians learn more about the electoral process, report on violations of electoral laws, and empower local partners to advocate for fair practices during election campaigning and voting. At a minimum, the project seems to have increased civic participation in election monitoring and raised political awareness more generally.

After applying his analytical framework and conducting interviews with DISC's U-Shahid users in Egypt, Meier concludes that while the project was not operating at a scale of visibility necessary to achieve significant governmental change, it "helped to reverse or at least fight back against this government-constructed panopticon, and this may have helped to pave the way for the 2011 revolution that toppled Mubarak." Moreover, U-Shahid was part of a broader landscape of intensive use of social media, including Facebook and blogs, "which filled the vacuum created by the lack of a real political debate in Egypt" and enabled young people "to engage in a political context in which physical elimination of the opposition was the norm."

One sees in Meier's case study as well the deficiencies of Egypt's authoritarian regime, which was significantly less adroit at using and constraining (or even fully understanding) these digital information and communication tools than its counterpart in China. As a result, it was caught flat-footed and outmaneuvered. This presaged to some extent the desperation of the Mubarak regime the following February, when its only response to the gathering, digitally driven crisis was to shut down Internet access altogether, a tactic which, it is generally now agreed, backfired.

In Chapter 8, Philip Howard and Muzammil Hussain come to bolder conclusions. They argue that digital media played a decisive role in the recent upheavals in Egypt and Tunisia. They find persuasive "the consistent narratives" from Arab civil society activists that "the Internet, mobile phones, and social media such as Facebook and Twitter made the difference this time. Using these technologies, people interested in democracy could build extensive networks, create social capital, and organize political action with a speed and on a scale never seen before." Each of the catalytic events of these revolutions was made so by liberation technology. When Mohamed Bouazizi, a Tunisian street vendor, set himself on fire after being humiliated by security officials, social media swiftly and powerfully publicized his horrific plight. What could have remained a local or individual tragedy was transformed into a national narrative of outrage by the viral cascade of YouTube videos, impassioned blogs, and SMS jokes about the decadent corruption and wanton abuses of President Zine al-Abidine Ben Ali's rule. When the Tunisian government moved to ban Facebook, Twitter, and YouTube, activists got support from international hacker communities to circumvent the state's firewalls, while also coordinating efforts mostly via SMS. Targeted arrests of bloggers did little to stem a fluid and leaderless revolution, which toppled Ben Ali on January 14, less than a month after the initial spark.

Campaigns of protest and civil disobedience quickly spread to a number of other Arab states, but Howard and Hussain stress that Egypt had one distinction that set it apart from the others: a more active and wired civil society. (Indeed, Egypt's April 6 Youth Movement had already used Facebook to organize a stunningly successful general strike in the spring of 2008.) Egypt's online public sphere was aroused by the rapidly escalating events in Tunisia as well as by the creation of a Facebook group, "We Are All Khaled Said," to sustain the memory of a young Egyptian who in June 2010 was dragged out of an Internet café and beaten to death by police after he exposed their corruption. That memorial page, launched by Google executive Wael Ghonim, "became a focus for collective dissent and commiseration," as well as "a logistical tool and, at least temporarily, a strong source of community."

Disaffected Egyptian youth readily identified with Said, "found solidarity through digital media, and then used their mobile phones to call their social networks into the streets." The breadth, speed, and discipline of protest mobilization disarmed both the regime and its domestic and international backers. When Mubarak tried to shut off Internet access in late January, it was too late. The shutdown was unevenly executed, and in any case it prompted many middle-class Egyptians who had heretofore stayed home to come into the streets. As digital media inspired, sustained, enlarged, and coordinated the peaceful protests, Mubarak's support base crumbled and he was forced to resign on 11 February 2011.

Elsewhere in the region, autocrats hung on in the face of similar digitally inspired protests, but in Bahrain and Syria it was only through extensive violent repression, and in Libya even that was not enough in the end. While Howard and Hussain give due weight to the underlying social and political sources of disaffection that destabilized Arab authoritarian regimes, they stress that "social discontent is not something ready-made" but must gestate and then be mobilized and organized into effective protest. In a number of Arab countries since late 2010, digital media have furnished the means for translating diffuse disenchantment "into workable strategies and goals," then a "structured movement with a collective consciousness," and then a mushrooming and nimble movement that "used Facebook, Twitter, and other sites to communicate plans for civic action." Mobile-phone photographs and videos uploaded to social media intensified domestic and international support for the protestors and raised the costs to the regime of blatant repression (though that has so far failed to restrain the Bashar al-Assad regime in Syria). Even the traditional media, such as the Arab satellite station Al Jazeera, relied heavily on digital media to collect information and images from countries in turmoil. Regimes fought back by trying to use or suppress these digital media, and in some instances (notably that of Saudi Arabia) they appeared to have the technological advantage over opposition forces from the very start.

Internet controls in the Arab world, and tools to help users circumvent them, form the subject that Walid Al-Saqaf addresses in Chapter 9. Al-Saqaf, a Yemeni journalist and software developer, is not only a scholar of online activity in the Arab world but also an activist and digital social entrepreneur. In 2007, he launched an online portal for his native Yemen that aggregated news, opinion articles, videos, blog posts, and other forms of online discussion. Although Yemen Portal brought together all sources of information—government, opposition, and independent media—it quickly became a "go-to" place for oppositional content, and almost as quickly the government built a firewall to block Internet access to it within Yemen. This prompted Al-Saqaf to develop a tool to evade censorship: "Alkasir," which means (literally) "the circumventor" in Arabic. Like similar tools, Alkasir works by having Internet "tunnels" access blocked websites through various proxy servers, so that the tunneled users can access the Internet as if they were connecting directly (save that the speed of connection can be much slower).

By January 2012, Alkasir had received more than a hundred-thousand reports of blocked websites in the Arab world. In engaging Arab netizens seeking this free software, Al-Saqaf was able to gather data that informs our understanding of the state of Internet censorship in Arab countries. Internet censorship, he finds, is "pervasive" and growing in many Arab countries. Not surprisingly, in addition to general social-media and multimedia-sharing sites such as Facebook and YouTube, the most frequently blocked sites contained independent news, dissident views, "criticism of the government, reports of human-rights violations, and similar content historically censored by autocratic governments in traditional media." And also not surprisingly, Syria and Saudi Arabia topped the list in terms of the number of web addresses reported blocked (along with Yemen, understandably, given his preexisting Internet constituency in that country).

While most analyses of Internet censorship are circumstantial and impressionistic, Al-Saqaf is able to present hard data demonstrating its scope and showing that Arab governments "have heavily invested in firewall software to suppress news and opinion content, social networks, and multimedia sharing." Yet he remains hopeful. By January of 2012, his software had been used more than three-million times by more than 30,000 users in Arab countries (two-thirds of them from Syria) and over 30,000 other users to access banned sites. He expects "the familiar cat-and-mouse game between ISPs in authoritarian Arab states" and the developers of circumvention tools to continue for years to come.

Mehdi Yahyanejad, the coauthor of our study of the use of social media by the Green Movement in Iran, also bridges the divide between analysis and Internet activism. He is the founder of the most popular Persian-language social news website, Balatarin.com. In Chapter 10, Yahyanejad and Elham Gheytanchi show how social-media tools such as blogs, Facebook, YouTube, Balatarin, and (to a lesser extent than was claimed at the time) Twitter were instrumental in informing Iranians and mobilizing protest following the disputed 2009 Iranian presidential election. Because they provided a secure and quick method for sharing news and information, "the new social media facilitated the exchange of ideas, fostered democratic political culture, and ultimately became an effective tool for mobilization" by the Green Movement.

In the years prior to the dramatic events of 2009—which saw the Islamic Republic conduct a deeply flawed presidential election and then suppress popular protest against the fraud—liberation technology tilled the soil for democratic change. In particular, "the Persian-speaking blogosphere," one of the ten largest in the world, raised the profile of opposition forces and "helped to create the discourse for democracy, pluralism, and tolerance in the years prior to the election." During the election campaign itself, the reformist presidential campaigns made intensive use of the Internet and social media, taking advantage of the regime's temporary unblocking of Facebook and Twitter, and mobilizing Iranian youth in unprecedented numbers.

Then, when politics shifted to the streets in protest of the stolen election, the widespread prevalence of ICTs in Iran and the relatively high level of Iranians' familiarity with digital tools (including circumvention tools) enabled protestors to use social media intensively to recruit their fellow citizens and to get the message out to the world, despite the government's physical and virtual crackdowns. A crucial dimension of this digital campaign was the uploading of video footage from mobile devices documenting the "the Iranian government's violent repression of its citizens' nonviolent protest movement." The iconic symbol for this brutality was the horrific image of an unarmed young woman, Neda Agha-Soltan, bleeding to death from gunshot wounds on a Tehran street. SMS and satellite-TV broadcasts were important supplements to web-based tools (with the BBC and Voice of America rebroadcasting YouTube content that was difficult to view online in Iran).

All these electronic media, along with creative nonviolent tactics, helped to sustain mass protests for months after the election campaign. But liberation technology could not, by itself, liberate Iran. The authoritarian regime, already adept at Internet censorship, showed that it, too, could learn and adapt. It used the same media to monitor tactical debates. Then it took back control of the streets and gradually arrested many of the protestors, using intercepted online communications to prosecute them. Yahyanejad and Gheytanchi take heart from the persistence of nonviolent resistance to the regime and remain hopeful about the role that social media can play in ultimately fostering democratic change in Iran. But they also recognize a sobering lesson: "Such media (with their open, horizontal nature) can also breed confusion when there is a need to deal with complex issues and tactics that require discipline, strategy, and a degree of central leadership." Iran's Green Movement, sadly, was deficient in all these respects.

In Chapter 11, Daniel Calingaert explains the opportunities and challenges for international policy to defend and promote Internet freedom. He is encouraged that "U.S. support for Internet freedom now is clearly articulated, identified as a priority in U.S. foreign policy, and backed by significant resources." He praises Secretary of State Hillary Clinton and the Obama Administration for pushing the policy agenda, supporting censorship circumvention and privacy-protection technologies, aiding digital activists, and integrating Internet freedom issues into the mainstream work of U.S. diplomacy. He believes, however, that the United States and the European Union both need to take stronger, more concerted action, because current policy "is inadequate to stem—let alone reverse—the trend of declining Internet freedom." Calingaert stresses what many of the authors in this volume have demonstrated—that authoritarian regimes such as those in China and Iran have become "highly adept at controlling the Internet." And he cites Freedom House reports showing that "threats to online freedom are growing and are increasingly diverse," as more countries begin to censor political content on the web and employ more tools and strategies for doing so.

To counteract these disturbing trends, Calingaert recommends a number of policy steps and approaches. First, the advanced democracies must do more to defend bloggers and cyberdissidents at risk in authoritarian regimes. Second, like other authors in this volume, he hopes that the United States will avoid civilian and military policy initiatives that might further fracture the Internet or legitimate authoritarian states' offensive and repressive actions on the web. These include the creation of technical "backdoors" to enable the U.S. government—but eventually also repressive governments—to intercept protected or encrypted messages. Third, he appeals for concrete assistance to U.S. (and European) media companies so that they may effectively respond to the urging of Secretary of State Clinton that they "take a proactive role in challenging foreign governments' demands for censorship and surveillance." He also views with favor banning exports of surveillance and censorship technology to highly authoritarian regimes, or at a minimum requiring

U.S. corporations to disclose requests from foreign governments to filter web content or hand over personal data of users. Fourth, he recommends challenging Internet censorship in appropriate international forums as both a human-rights issue and a restriction on free trade. Fifth, he seeks "targeted diplomatic initiatives to challenge restrictive Internet laws and practices" before they become entrenched. And finally, he urges the major democracies to act in concert to promote and protect digital freedom and to coordinate their assistance policies.

Liberation or Control?

It is far too soon to know whether digital ICTs will come to be seen more as instruments of liberation or as tools for social and political control. Even in liberal democracies, there are concerns that citizens are permanently losing their privacy to both state and corporate actors, and that elected governments are subverting liberal norms of constitutionalism and individual rights in a headlong rush for technological advantage in the war on terrorism, the war on cybercrime, and indeed the reconceptualization of war itself. In authoritarian regimes, as this volume shows, the balance of technological prowess does matter; in Egypt and Tunisia it seemed to favor youthful democratic oppositions. Yet these two regimes, and those in Bahrain, Libya, Syria, and Yemen, suffered a crisis of legitimacy beyond anything that Chinese Communist rule now faces. Should China's Communist Party lose "the Mandate of Heaven" as thoroughly as Tunisia's Ben Ali and Egypt's Mubarak had lost the support of their respective peoples, it is unlikely that it will be saved merely by the breadth and sophistication of its Internet controls. Just as the army defected or stood aside in Tunisia and Egypt, so the loyalty of China's vast network of Internet police and information censors might also waver catastrophically in a crisis. Even if China has gradually constructed a more diffuse, subtle, and multilayered architecture of digital control, that system is only as effective as the people who maintain it. As I argue in the first chapter, "It is not technology, but people, organizations, and governments that will determine who prevails."

In the words of an Egyptian digital activist interviewed by Patrick Meier, it is possible to find grounds for hope: "Technology by nature is a very neutral tool. But the most important thing is information. Information is the key that drives political discourse and media debates. Information wants to be found. Those who want to suppress it will have a harder time. So people in favor of spreading information are going to win."

NOTES

1. See *www.benphoster.com/facebook-user-growth-chart-2004-2010.*

2. See *http://blog.twitter.com/2010/02/measuring-tweets.html; http://blog.twitter.com /2011/06/200-million-tweets-per-day.html.*

3. See *www.website-monitoring.com/blog/2010/05/17/youtube-facts-and-figures-history-statistics.*

4. Peter Diamandis and Steven Kotler, *Abundance: The Future Is Better Than You Think* (New York: Free Press, 2012).

5. "Buzz in the Blogosphere: Millions More Bloggers and Blog Readers," Nielsenwire, 8 March 2012, *http://blog.nielsen.com/nielsenwire/online_mobile/buzz-in-the-blogosphere -millions-more-bloggers-and-blog-readers.*

6. Larry Diamond, *The Spirit of Democracy: The Struggle to Build Free Societies Throughout the World* (New York: Times Books, 2008), 340–44.

7. A useful gateway to and clearinghouse for innovations in ICT4D can be found at the website of Infodev, a partnership program of the World Bank Group, *www.infodev.org.*

8. See *http://opennet.net.*

9. As of mid-2011, China had an estimated 485 million Internet users: *www.inter-networldstats.com/top20.htm.* However, in its January 2012 Statistical Report, the China Network Information Center, the official Chinese body tracking Internet use, put the number of users at 513 million, as of the end of 2011: *www1.cnnic.cn/uploadfiles/ pdf/2012/2/27/112543.pdf.* This is also the source for the number of mobile-phone users. These are all estimates based on sample surveys.

10. See *www.politifact.com/truth-o-meter/statements/2011/aug/19/jon-huntsman/jon-huntsman-says-internet-use-china-forum-discon.*

I

Liberation vs. Control in Cyberspace

1

LIBERATION TECHNOLOGY

Larry Diamond

Larry Diamond *is senior fellow at the Hoover Institution and the Freeman Spogli Institute for International Studies, director of Stanford University's Center on Democracy, Development, and the Rule of Law, and founding coeditor of the* Journal of Democracy. *This essay originally appeared in the July 2010 issue of the* Journal of Democracy.

In March 2003, police in Guangzhou (Canton), China, stopped 27-year-old Sun Zhigang and demanded to see his temporary living permit and identification. When he could not produce these, he was sent to a detention center. Three days later, he died in its infirmary. The cause of death was recorded as a heart attack, but the autopsy authorized by his parents showed that he had been subjected to a brutal beating.

Sun's parents took his story to the liberal newspaper *Nanfang Dushi Bao* (Southern Metropolis Daily), and its investigation confirmed that Sun had been beaten to death in custody. As soon as its report appeared on April 25, "newspapers and Web sites throughout China republished the account, [Internet] chat rooms and bulletin boards exploded with outrage," and it quickly became a national story.[1] The central government was forced to launch its own investigation and on June 27, it found twelve people guilty of Sun's death.

Sun's case was a rare instance in China of official wrongdoing being exposed and punished. But it had a much wider and more lasting impact, provoking national debate about the "Custody and Repatriation" (C&R) measures that allowed the police to detain rural migrants (typically in appalling conditions) for lacking a residency or temporary-living permit. In the outrage following Sun's death, numerous Chinese citizens posted on the Internet stories of their own experiences of C&R, and the constitutionality of the legislation became a hotly debated topic in universities. An online petition asking the Standing Committee of the National People's Congress to reexamine C&R quickly garnered widespread popular support, and in June 2003 the government announced that it would close all of the more than eight-hundred C&R detention centers.[2]

Sun's case was seen as a watershed—the first time that a peaceful outpouring of public opinion had forced the Communist Chinese state to change a national regulation. But Sun's case also soon became that of muckraking editor Cheng Yizhong, whom local officials jailed (along with three of his colleagues) in retaliation for their efforts to ferret out the wrongdoing that led to Sun's death. The legal defense that Xu Zhiyong mounted on behalf of the four journalists itself became a *cause célèbre*. As their fellow journalists launched an unprecedented campaign for their release, using among other means an Internet petition, Xu established a website, the Open Constitutional Initiative, to post documents and legal arguments about the case. All of this reflected a burgeoning *weiquan* ("defend-rights") movement. But while Cheng and his deputy editor were released from prison without charge, they lost their jobs and the authorities closed down Xu's site. Xu continued his work in defense of rights until July of last year, when his organization was shut down and he was arrested on politically motivated charges of tax evasion.

Optimists discern in these events a striking ability of the Internet—and other forms of "liberation technology"—to empower individuals, facilitate independent communication and mobilization, and strengthen an emergent civil society. Pessimists argue that nothing in China has fundamentally changed. The Chinese Communist Party (CCP) remains firmly in control and beyond accountability. The *weiquan* movement has been crushed. And the Chinese state has developed an unparalleled system of digital censorship.

Both perspectives have merit. Liberation technology enables citizens to report news, expose wrongdoing, express opinions, mobilize protest, monitor elections, scrutinize government, deepen participation, and expand the horizons of freedom. But authoritarian states such as China, Belarus, and Iran have acquired (and shared) impressive technical capabilities to filter and control the Internet, and to identify and punish dissenters. Democrats and autocrats now compete to master these technologies. Ultimately, however, not just technology but political organization and strategy and deep-rooted normative, social, and economic forces will determine who "wins" the race.

Liberation technology is any form of information and communication technology (ICT) that can expand political, social, and economic freedom. In the contemporary era, it means essentially the modern, interrelated forms of digital ICTs—the computer, the Internet, the mobile phone, and countless innovative applications for them, including "new social media" such as Facebook and Twitter. Digital ICTs have some exciting advantages over earlier technologies. The Internet's decentralized character and ability (along with mobile-phone networks) to reach large numbers of people very quickly, are well suited to grassroots organizing. In sharp contrast to radio and television, the new ICTs are two-way and even multiway forms of communication. With tools such as Twitter (a social-networking and microblogging service allowing its users to send and read messages with up to 140 characters), a user can instantly reach hundreds or even thousands of "followers." Users are thus not just passive recipients but journalists, commentators, videographers, entertainers, and organizers. Although most of this use is not political, the technology can empower those who wish to become political and to challenge authoritarian rule.

It is tempting to think of the Internet as unprecedented in its potential for political progress. History, however, cautions against such hubris. In the fifteenth century, the printing press revolutionized the accumulation and dissemination of information, enabling the Renaissance, the Protestant Reformation, and the scientific revolution. On these foundations, modern democracy emerged. But the printing press also facilitated the rise of the centralized state and prompted the movement toward censorship.[3] A century and a half ago, the telegraph was hailed as a tool to promote peace and understanding. Suddenly, the world shrank; news that once took weeks to travel across the world could be conveyed instantly. What followed was not peace and freedom but the bloodiest century in human history. Today's enthusiasts of liberation technology could be accused of committing the analytic sins of their Victorian forebears, "technological utopianism" and "chronocentricity"—that is, "the egotism that one's own generation is poised on the very cusp of history."[4]

In the end, technology is merely a tool, open to both noble and nefarious purposes. Just as radio and TV could be vehicles of information pluralism and rational debate, so they could also be commandeered by totalitarian regimes for fanatical mobilization and total state control. Authoritarian states could commandeer digital ICTs to a similar effect. Yet to the extent that innovative citizens can improve and better use these tools, they can bring authoritarianism down—as in several cases they have.

Mobilizing against authoritarian rule represents only one possible "liberating" use of digital ICTs. Well before mobilization for democracy peaks, these tools may help to widen the public sphere, creating a more pluralistic and autonomous arena of news, commentary, and information. The new ICTs are also powerful instruments for transparency and accountability, documenting and deterring abuses of human rights and democratic procedures. And though I cannot elaborate here, digital ICTs are also liberating people from poverty and ill health: conveying timely information about crop prices, facilitating microfinance for small entrepreneurs, mapping the outbreaks of epidemics, and putting primary healthcare providers in more efficient contact with rural areas.[5]

Malaysia: Widening the Public Sphere

A crucial pillar of authoritarian rule is control of information. Through blogs (there are currently more than a hundred million worldwide), blog sites, online chatrooms, and more formal online media, the Internet provides dramatic new possibilities for pluralizing flows of information and widening the scope of commentary, debate, and dissent.

One of the most successful instances of the latter type is *Malaysiakini*, an online .newspaper that has become Malaysia's principal alternative source of news and commentary.[6] As Freedom House has documented, Malaysia lacks freedom of the press. The regime (both the state and the ruling Barisan Nasional [BN] coalition) dominates print and broadcast media through direct ownership and monopoly practices. Thus it can shape what Malaysians read and see, and it can punish critical journalists with dismissal. Repressive laws severely constrain freedom to report, publish, and broadcast. However, as a rapidly developing country with high literacy, Malaysia has witnessed explosive growth of Internet access (and recently, broadband access), from 15 percent of the population in 2000 to 66 percent in 2009 (equal to Taiwan and only slightly behind Hong Kong).[7] The combination of tight government control of the conventional media, widespread Internet access, and relative freedom on the Internet created an opening for online journalism in Malaysia, and two independent journalists—Steven Gan and Premesh Chandran—ventured into it. Opponents of authoritarian rule since their student days, Gan and Chandran became seized during the 1998 *reformasi* period with the need to reform the media and bring independent news and reporting to Malaysia. Using about US$9,000 of their own money (a tiny fraction of what it would take to start a print newspaper), they launched *Malaysiakini* in November 1999. Almost immediately, they gained fame by exposing how an establishment newspaper had digitally cropped jailed opposition leader (and former deputy prime minister) Anwar Ibrahim from a group photo of ruling-party politicians.

From its inception, *Malaysiakini* has won a loyal and growing readership by providing credible, independent reporting on Malaysian politics and governance. As its readership soared, that of the mainstream newspapers fell. Suddenly, Malaysians were able to read about such long-taboo subjects as corruption, human-rights abuses, ethnic discrimination, and police brutality. Now the online paper posts in English about fifteen news stories a day, in addition to opinion pieces, letters, readers' comments, and daily satire (in *Cartoonkini*), plus translations and original material in Chinese, Malay, and Tamil. *Malaysiakini* reports scandals that no establishment paper would touch, such as massive cost overruns related to conflicts of interest at the country's main port agency and ongoing financial misconduct at the government-supported Bank Islam Malaysia. With the regime's renewed legal assault on Anwar Ibrahim, *Malaysiakini* is the only place where Malaysians can turn for independent reporting on the legal persecution of the opposition leader. In July 2008, it became Malaysia's most visited news site with about 2.5 million visitors per month. Yet, like many online publications worldwide, it still strives for financial viability.

While Malaysia today is no less authoritarian than when *Malaysiakini* began publishing a decade ago, it is more competitive and possibly closer to a democratic breakthrough than at any time in the last four decades. If a transition occurs, it will be mainly due to political factors—the coalescence of an effective opposition and the blunders of an arrogant regime. In addition, economic and social change is generating a better-educated and more diverse population, less tolerant of government paternalism and control. Polling and other data show that young Malaysians in particular support the (more democratic) opposition. But it is hard to disentangle these political and social factors from the expansion of the independent public sphere that *Malaysiakini* has spearheaded. In March 2008, the BN made its worst showing at the polls in half a century, losing its two-thirds parliamentary majority for the first time since independence. Facilitating this was the growing prominence of online journalism, which diminished the massive BN advantage in media access and "shocked the country" by documenting gross police abuse of demonstrators, particularly those of Indian descent.

Malaysiakini and its brethren perform a number of democratic functions. They report news and convey images that Malaysians would not otherwise see. They provide an uncensored forum for commentary and debate, giving rise to a critical public sphere. They offer space and voice to those whose income, ethnicity, or age put them on the margins of society. They give the political opposition, which is largely shut out of the establishment media, a chance to make its case. In the process, they educate Malaysians politically and foster more democratic norms. Many online publications and Internet blog sites perform similar functions in other semi-authoritarian countries, such as Nigeria, and in emerging and illiberal democracies. But is it possible for these functions to take root in a country as authoritarian as China is today?

Opening a Public Sphere in China

The prevailing answer is no: China's "Great Firewall" of Internet filtering and control prevents the rise of an independent public sphere online. Indeed, China's policing of the Internet is extraordinary in both scope and sophistication. China now has the world's largest population of Internet users—more than 380 million people (a number equal to 29 percent of the population, and a sixteen-fold increase since the year 2000). But it also has the world's most extensive, "multilayered," and sophisticated system "for censoring, monitoring, and controlling activities on the internet and mobile phones."[8] Connection to the international Internet is monopolized by a handful of state-run operators hemmed in by rigid constraints that produce in essence "a national intranet," cut off from anything that might challenge the CCP's monopoly on power.

Access to critical websites and online reporting is systematically blocked. Google has withdrawn from China in protest of censorship, while YouTube, Facebook, and Blogspot, among other widely used sites, are extensively blocked or obstructed. Chinese companies that provide search and networking services agree to even tighter self-censorship than do international companies. When protests erupt (as they did over Tibet in 2008, for instance) or other sensitive political moments approach, authorities preemptively close data centers and online forums. Now the party-state is also trying to eliminate anonymous communication and networking, by requiring registration of real names to blog or comment and by tightly controlling and monitoring cybercafés. Fifty-thousand Internet police prowl cyberspace removing "harmful content"—usually within 24 to 48 hours. Students are recruited to spy on their fellows. And the regime pays a quarter of a million online hacks (called "50-centers" because of the low piece rate they get) to post favorable comments about the party-state and report negative comments.

Such quasi-Orwellian control of cyberspace is only part of the story, however. There is simply too much communication and networking online (and via mobile phones) for the state to monitor and censor it all. Moreover, Chinese "netizens"—particularly the young who are growing up immersed in this technology—are inventive, determined, and cynical about official orthodoxy. Many constantly search for better techniques to circumvent cybercensorship, and they quickly share what they learn. If most of China's young Internet users are apolitical and cautious, they are also alienated from political authority and eagerly embrace modest forms of defiance, often turning on wordplay.

Recently, young Chinese bloggers have invented and extensively lauded a cartoon creature they call the "grass mud horse" (the name in Chinese is an obscene pun) as a vehicle for protest. This mythical equine, so the narrative goes, is a brave and intelligent animal whose habitat is threatened by encroaching "river crabs." In Chinese, the name for these freshwater crustaceans *(hexie)* sounds very much like the word for Hu Jintao's official governing philosophy of "harmony"—a label that critics see as little more than a euphemism for censorship and the suppression of criticism. Xiao Qiang, editor of *China Digital Times,* argues that the grass mud horse

> has become an icon of resistance to censorship. The expression and cartoon videos may seem like a juvenile response to unreasonable rule. But the fact that the vast online population has joined the chorus, from serious scholars to usually politically apathetic urban white-collar workers, shows how strongly this expression resonates.[9]

In order to spread defiance, Chinese have a growing array of digital tools. Twitter has become one of the most potent means for political and social networking and the rapid dissemination of news, views, and withering satire. On April 22 at People's University in Beijing, three human-rights activists protested a speech by a well-known CCP propaganda official, Wu Hao. Showering him with small bills, they declared, "Wu Hao, wu mao!" ("Wu Hao is a fifty-center!"). Twitter flashed photographs of the episode across China, delighting millions of students who revel in mocking the outmoded substance, tortured logic, and painfully crude style of regime propagandists.

When Google announced in late March 2010 it was withdrawing its online search services from mainland China (after failing to resolve its conflict with the government over censorship and cyberattacks), the Chinese Twitter-sphere lit up. Many Chinese were upset that Google would abandon them to the more pervasive censorship of the Chinese search-engine alternatives (such as Baidu), and they worried that the Great Firewall would block other services such as Google Scholar and Google Maps. Others suspected Google of doing the U.S. government's bidding. But the company's decision provoked a wave of sympathy and mourning, similar to what happened in January when Google first announced that it was considering withdrawing: "Citizen reporters posted constant updates on . . . Twitter, documenting the Chinese netizens who endlessly offered flowers, cards, poems, candles, and even formal bows in front of the big outdoor sign 'Google' located outside the company's offices in Beijing, Shanghai and Guangzhou."[10] Security guards chased the mourners away, declaring the offerings "illegal flower tributes." The term quickly spread in China's online forums, symbolizing the suppression of freedom.

The public sphere in China involves much more than "tweets," of course. Those often link to much longer blogs, discussion groups, and news reports. And many thought-provoking sites are harder to block because their critiques of CCP orthodoxy are subtler, elucidating democratic principles and general philosophical concepts, sometimes with reference to Confucianism, Taoism, and other strains of traditional Chinese thought that the CCP dares not ban. Full-scale blog posts (not subject to Twitter's severe length limits) are far likelier to criticize the government (albeit artfully and euphemistically). Rebecca MacKinnon finds that China's blogosphere is a "much more freewheeling space than the mainstream media," with censorship varying widely across the fifteen blog-service providers that she examined. Thus, "a great deal of politically sensitive material survives in the Chinese blogosphere, and chances for survival can likely be improved with knowledge and strategy."[11]

Despite the diffuse controls, China's activists see digital tools such as Twitter, Gmail, and filtration-evading software as enabling levels of communication, networking, and publishing that would otherwise be unimaginable in China today. With the aid of liberation technology, dissident intellectuals have gone from being a loose assortment of individuals with no specific goal or program to forming a vibrant and increasingly visible collaborative force. Their groundbreaking manifesto—Charter 08, a call for nineteen reforms to achieve "liberties, democracy, and the rule of law" in China—garnered most of its signatures through the aid of blog sites such as *bullog.cn*. When Charter 08 was released online on 10 December 2008, with the signatures of more than three-hundred Chinese intellectuals and human-rights activists, the government quickly moved to suppress all mention of it. But then, "something unusual happened. Ordinary people such as Tang [Xiaozhao] with no history of challenging the government began to circulate the document and declare themselves supporters," shedding their previous fear. Within a month, more than five-thousand other Chinese citizens had signed the document. They included not just the usual dissidents but "scholars, journalists, computer technicians, businessmen, teachers and students whose names had not been associated with such movements before, as well as some on the lower rungs of China's social hierarchy—factory and construction workers and farmers."[12]

Officials shut down Tang's blog soon after she signed the Charter, and did the same to countless other blogs that supported it (including the entire *bullog.cn* site). But the campaign persists in underground salons, elliptical references, and subversive jokes spread virally through social media and instant messaging. One such joke imagines a testy Chinese president Hu Jintao complaining about the Charter's democratic concepts such as federalism, opposition parties, and freedom of association. "Where do they all come from?" he demands. His minions run down the sources and bring him the bad news: The troublesome notions can be traced to Mao Zedong, Zhou Enlai, the CCP, the official newspaper (the *Xinhua Daily*), and the constitution of the People's Republic itself. A flustered Hu wonders what to do. His staff suggests banning all mention of these names. "You idiots!" shouts Hu. "If you ban them, you might as well ban me too!" "Well," his staff retorts, "People do say that if they ban you, at least the Charter will be left alone."[13]

Monitoring Governance, Exposing Abuses

Liberation technology is also "accountability technology," in that it provides efficient and powerful tools for transparency and monitoring. Digital cameras combined with sites such as YouTube create new possibilities for exposing and challenging abuses of power. Incidents of police brutality have been filmed on cellphone cameras and posted to YouTube and other sites, after which bloggers have called outraged public attention to them. Enter "human rights abuses" into YouTube's search box and you will get roughly ten-thousand videos showing everything from cotton-growers' working conditions in Uzbekistan, to mining practices in the Philippines, to human-organ harvesting in China, to the persecution of Bahá'ís in Iran. A YouTube video of a young Malaysian woman forced by the police to do squats while naked forced the country's prime minister to call for an independent inquiry. When Venezuelan president Hugo Chávez forced Radio

Caracas Television off the air in May 2007, it continued its broadcasts via YouTube. No wonder, then, that authoritarian states such as Iran and Saudi Arabia completely block access to that video-posting site.

Across much of the world, and especially in Africa, the quest for accountability makes use of the simplest form of liberation technology: text messaging via mobile phone. (Mobile-phone networks have proven particularly useful in infrastructure-starved Africa since they can cover vast areas without requiring much in the way of physical facilities beyond some cell towers.) Around the world, the reach and capabilities of cellphones are being dramatically expanded by open-source software such as FrontlineSMS, which enables large-scale, two-way text messaging purely via mobile phones. In recent years, the software has been used over mobile-phone networks to monitor national elections in Nigeria and Ghana, to facilitate rapid reporting of human-rights violations in Egypt, to inform citizens about anticorruption and human-rights issues in Senegal, and to monitor and report civil unrest in Pakistan. A Kenyan organization, Ushahidi (Swahili for "testimony"), has adapted the software for "crisis-mapping." This allows anyone to submit crisis information through text messaging using a mobile phone, e-mail, or online-contact form, and then aggregates the information and projects it onto a map in real time. It was initially developed by citizen journalists to map reports of postelection violence in Kenya in early 2008, drawing some 45,000 Kenyan users. It has since been used to report incidents of xenophobic violence in South Africa; to track violence and human-rights violations in the Democratic Republic of Congo; and to monitor elections in Afghanistan, India, Lebanon, and Mexico.

The largest funder of both Ushahidi and FrontlineSMS is the Omidyar Network (ON), a philanthropic investment firm established six years ago by eBay founder Pierre Omidyar and his wife Pam. It extends into the worlds of political and social innovation the eBay approach: giving everyone equal access to information and opportunity to leverage the potential of individuals and the power of markets. This innovative effort—which comprises both a venture-capital fund directed at for-profit start-ups and a nonprofit grant-making fund—has committed more than $325 million in investments and grants in two broad areas: "access to capital" (microfinance, entrepreneurship, and property rights), and "media, markets and transparency" (which supports technology that promotes transparency, accountability, and trust across media, markets, and government). The ON supports national partners in Nigeria, Ghana, and Kenya that are using information technology to improve governance and free expression. These include Infonet—a web portal that provides citizens, media, and NGOs with easy-to-access information on national- and local-government budgets in Kenya—and Mzalendo, a comprehensive site that enables Kenyans to follow what their members of parliament are doing.

The ON's support for transparency initiatives also extends to other countries and to U.S.-based organizations. These include Global Integrity, which harnesses the Internet and other sources of information in order to generate detailed assessments of corruption in more than ninety countries; and the Sunlight Foundation, which utilizes the Internet and related technology in order to make information about federal-government spending, legislation, and decision making more accessible to U.S. voters.

Mobilizing Digitally

One of the most direct, powerful, and—to authoritarian regimes—alarming effects of the digital revolution has been its facilitation of fast, large-scale popular mobilizations. Cellphones with SMS text messaging have made possible what technology guru Howard Rheingold calls "smart mobs"—vast networks of individuals who communicate rapidly and with little hierarchy or central direction in order to gather (or "swarm") at a certain location for the sake of protest. In January 2001, Philippine president Joseph Estrada "became the first head of state in history to lose power to a smart mob," when tens of thousands and then, within four days, more than a million digitally mobilized Filipinos assembled at a historic protest site in Manila.[14] Since then, liberation technology has been instrumental in virtually all of the instances where people have turned out *en masse* for democracy or political reform.

Liberation technology figured prominently in the Orange Revolution that toppled the electoral authoritarian regime in Ukraine via mass protests during November and December 2004. The Internet newspaper *Ukrainskaya Pravda* provided a vital source of news and information about both the regime's efforts to steal the presidential election and the opposition's attempts to stop it. By the revolution's end, this online paper had become "the most widely read news source of any kind in Ukraine."[15] Website discussion boards gave activists a venue for documenting fraud and sharing best practices.[16] Text messaging helped to mobilize and coordinate the massive public protests—bringing hundreds of thousands to Kyiv's Independence Square in freezing weather—that ultimately forced a new runoff, won by the democratic opposition.

These digital tools also facilitated the 2005 Cedar Revolution in Lebanon (which drew more than a million demonstrators to demand the withdrawal of Syrian troops); the 2005 protests for women's voting rights in Kuwait; the 2007 protests by Venezuelan students against the closure of Radio Caracas Television; and the April 2008 general strike in Egypt, where tens of thousands of young demonstrators mobilized through Facebook.[17] In September 2007, the "Internet, camera phones, and other digital networked technologies played a critical role" in Burma's Saffron Revolution, so called because of the involvement of thousands of Buddhist monks. Although digital technology did little directly to mobilize the protests, it vividly informed the world of them, and revealed the bloody crackdown that the government launched in response: "Burmese citizens took pictures and videos, many on their mobile phones, and secretly uploaded them from Internet cafes or sent digital files across the border to be uploaded." This international visibility may have saved many lives by inhibiting the military from using force as widely and brutally as it had in 1988.[18]

In China, pervasive text messaging has been a key factor in the mushrooming of grassroots protests. In 2007, an eruption of hundreds of thousands of cellphone text messages in Xiamen, a city on the Taiwan Strait, generated so much public dismay at the building of an environmentally hazardous chemical plant that authorities suspended the project.[19] The impact of the text messages was magnified and spread nationally as bloggers in other Chinese cities received them and quickly fanned the outrage. The technology is even seeping into North Korea, the world's most closed society, as North Korean defectors and South Korean human-rights activists entice North Koreans to carry the phones back home with them from China and then use them to report what is happening (via the Chinese mobile network).[20] In the oil-rich Gulf states, text messaging allows civic activists and political oppositionists "to build unofficial membership lists, spread news about detained activists, encourage voter turnout, schedule meetings and rallies, and develop new issue campaigns—all while avoiding government-censored newspapers, television stations, and Web sites."[21]

The most dramatic recent instance of digital mobilization was Iran's Green Movement, following the egregious electoral malpractices that appeared to rob opposition presidential candidate Mir Hosein Musavi of victory on 12 June 2009. In the preceding years, Iran's online public sphere had been growing dramatically, as evidenced by its more than "60,000 routinely updated blogs" exploring a wide range of social, cultural, religious, and political issues;[22] the explosion of Facebook to encompass an estimated 600,000 Persian-language users;[23] and the growing utilization of the Internet by news organizations, civic groups, political parties, and candidates.

As incumbent president Mahmoud Ahmedinejad's election victory was announced (complete with claims of a 62 percent landslide) on June 13, outraged accounts of vote fraud spread rapidly via Internet chatrooms, blogs, and social networks. Through Twitter, text messaging, Facebook, and Persian-language social-networking sites such as Balatarin and Donbleh, Iranians quickly spread news, opinions, and calls for demonstrations. On June 17, Musavi supporters used Twitter to attract tens of thousands of their fellow citizens to a rally in downtown Tehran. Internet users organized nationwide protests throughout the month, including more large demonstrations in the capital, some apparently attended by two to three million people. YouTube also provided a space to post pictures and videos of human-rights abuses and government crackdowns. A 37-second video of the death of Neda Agha-Soltan during Tehran's violent protests on

June 20 quickly spread across the Internet, as did other images of the police and regime thugs beating peaceful demonstrators. Neda's death and the distressing images of wanton brutality decimated the remaining legitimacy of the Islamic Republic domestically and internationally.

To date, the Green Movement illustrates both the potential and limits of liberation technology. So far, the Islamic Republic's reactionary establishment has clung to power through its control over the instruments of coercion and its willingness to wield them with murderous resolve. Digital technology could not stop bullets and clubs in 2009, and it has not prevented the rape, torture, and execution of many protestors. But it has vividly documented these abuses, alienating key pillars of the regime's support base, including large segments of the Shia clergy. While the regime has tortured dissidents to get their e-mail passwords and round up more opponents, the Internet has fostered civic and political pluralism in Iran; linked the opposition within that country to the Iranian diaspora and other global communities; and generated the consciousness, knowledge, and mobilizational capacity that will eventually bring down autocracy in Iran. A key factor affecting when that will happen will be the ability of Iranians to communicate more freely and securely online.

Breaking Down the Walls

Even in the freest environments, the new digital means of information and communication have important limits and costs. There are fine lines between pluralism and cacophony, between advocacy and intolerance, and between the expansion of the public sphere and its hopeless fragmentation. As the sheer number of media portals has multiplied, more voices have become empowered, but they are hardly all rational and civil. The proliferation of online (and cable) media has not uniformly improved the quality of public deliberation, but rather has given rise to an "echo chamber" of the ideologically like-minded egging each other on. And open access facilitates much worse: hate-mongering, pornography, terrorism, digital crime, online espionage, and cyberwarfare. These are real challenges, and they require careful analysis—prior to regulation and legislation—to determine how democracies can balance the great possibilities for expanding human freedom, knowledge, and capacity with the dangers that these technologies may pose for individual and collective security alike.

Still the overriding challenge for the digital world remains freedom of access. The use of Internet filtering and surveillance by undemocratic regimes is becoming both more widespread and more sophisticated. And some less-sophisticated efforts, using commercial filtering software, may block sites even more indiscriminately. Currently, more than three-dozen states filter the Internet or completely deny their citizens access.[24] Enterprising users can avail themselves of many circumvention technologies, but some require installation of software and so will not be available if the Internet is accessed from public computers or Internet cafes; many of the Web-based applications are blocked by the same filters that block politically sensitive sites; and most of these means require some degree of technical competence by the user.[25] Not all circumvention methods protect netizens' privacy and anonymity, which can be a particularly acute problem when state-run companies provide the Internet service. The free software Tor, popular among Iranians, promises anonymity by "redirecting encrypted traffic through multiple relays . . . around the world," making it difficult for a regime to intercept a transmission.[26] But if it effectively monopolizes the provision of Internet service, a desperate regime such as Burma's in 2007 can always respond by shutting down the country's Internet service or, as Iran's government did, by slowing service to a paralyzing crawl while authorities searched electronic-data traffic for protest-related content.[27]

Even in liberal democracies, issues of access arise. Recently netizens worldwide—and the U.S. government—have become concerned over excessively broad legislative proposals in Australia that would force Internet service providers to blacklist a large number of sites for legal and moral considerations (including the protection of children). The Chinese practice of forcing Internet providers to assume liability for the content to which they provide access is seeping into European legal and regulatory thinking regarding the Internet.[28]

There is now a technological race underway between democrats seeking to circumvent Internet censorship and dictatorships that want to extend and refine it. Recently, dictatorships such as Iran's have made significant gains in repression. In part, this has happened because Western companies like Nokia-Siemens are willing to sell them advanced surveillance and filtering technologies. In part, it has also been the work of dictatorships that eagerly share their worst practices with one another. A host of new circumvention technologies are coming onto the market, and millions of Chinese, Vietnamese, Iranians, Tunisians, and others fervently want access to them. Rich liberal democracies need to do much more to support the development of such technologies, and to facilitate (and subsidize) their cheap and safe dissemination to countries where the Internet is suppressed. More could be done to improve encryption so that people in authoritarian regimes can more safely communicate and organize online. Breakthroughs may also come with the expansion of satellite access that bypasses national systems, if the cost of the satellite dishes and monthly usage rates can be reduced dramatically. Western governments can help by banning the export of advanced filtering and surveillance technologies to repressive governments, and by standing behind Western technology companies when dictatorships pressure them "to hand over Internet users' personal data."[29] And finally, liberal democracies should stand up for the human rights of bloggers, activists, and journalists who have been arrested for peacefully reporting, networking, and organizing online.

It is important for the United States to have declared, as Secretary of State Hillary Clinton did in a historic speech on 21 January 2010, that "We stand for a single Internet where all of humanity has equal access to knowledge and ideas." But the struggle for electronic access is really just the timeless struggle for freedom by new means. It is not technology, but people, organizations, and governments that will determine who prevails.

NOTES

The author thanks Anna Davies, Blake Miller, and Astasia Myers for their truly superb research assistance on this article; and also Lian Matias, Galen Panger, Tucker Herbert, Ryan Delaney, Daniel Holleb, Sampath Jinadasa, and Aaron Qayumi for their prior research assistance on this project.

1. Sophie Beach, "The Rise of Rights?" *China Digital Times, http://chinadigitaltimes. net/2005/05/rise-of-rights.*

2. Yongnian Zheng, *Technological Empowerment: The Internet State and Society in China* (Stanford: Stanford University Press, 2008), 147–51.

3. Ithiel de Sola Pool, *Technologies of Freedom* (Cambridge, Mass.: Belknap Press, 1983), 251.

4. Tom Standage, *The Victorian Internet* (New York: Berkley, 1998), 210, 213.

5. For various accounts, see *http://fsi.stanford.edu/research/program_on_liberation_ technology.*

6. This account draws heavily from a student research paper conducted under my supervision: Astasia Myers, "*Malaysiakini:* Internet Journalism and Democracy," Stanford University, 4 June 2009.

7. Figures on the growth of Web use in Malaysia and China are available at *www.internetworldstats.com/stats3.htm.*

8. Freedom House, "Freedom on the Net: A Global Assessment of Internet and Digital Media," 1 April 2009, 34; available at *www.freedomhouse.org.*

9. Private email message from Xiao Qiang, May 2009. Quoted with permission.

10. S.L. Shen, "Chinese Forbidden from Presenting Flowers to Google," UPI Asia Online, 15 January 2010; available at *www.upiasia.com/Politics/2010/01/15/chinese_forbidden_from_presenting_flowers_to_google/4148.*

11. Rebecca MacKinnon, "China's Censorship 2.0: How Companies Censor Bloggers," *First Monday,* 2 February 2009; available at *http://firstmonday.org/htbin/cgiwrap/bin/ojs/index.php/fm/article/view/2378/2089.* See also Ashley Esarey and Xiao Qiang, "Below the Radar: Political Expression in the Chinese Blogosphere," *Asian Survey* 48 (September–October 2008): 752–72.

12. Ariana Eunjung Cha, "In China, a Grass-Roots Rebellion," *Washington Post,* 29 January 2009.

13. "Charter 08 Still Alive in the Chinese Blogosphere," *China Digital Times,* 9 February 2009.

14. Howard Rheingold, *Smart Mobs: The Next Social Revolution* (New York: Basic Books, 2003), 158.

15. Michael McFaul, "Transitions from Postcommunism," *Journal of Democracy* 16 (July 2005): 12.

16. Robert Faris and Bruce Etling, "Madison and the Smart Mob: The Promise and Limitations of the Internet for Democracy," *Fletcher Forum of World Affairs* 32 (Summer 2008): 65.

17. Cathy Hong, "New Political Tool: Text Messaging," *Christian Science Monitor*, 30 June 2005; Jóse de Córdoba, "A Bid to Ease Chávez's Power Grip; Students Continue Protests in Venezuela; President Threatens Violence," *Wall Street Journal*, 8 June 2007.

18. Mridul Chowdhury, "The Role of the Internet in Burma's Saffron Revolution," Berkman Center for Internet and Society, September 2008, 14 and 4; available at *http://cyber.law.harvard.edu/sites/cyber.law.harvard.edu/files/Chowdhury_Role_of_the_Internet_in_Burmas_Saffron_Revolution.pdf_0.pdf*.

19. Edward Cody, "Text Messages Giving Voice to Chinese," *Washington Post*, 28 June 2007.

20. Choe Sang-Hun, "North Koreans Use Cell Phones to Bare Secrets," *New York Times*, 28 March 2010. Available at *www.nytimes.com/2010/03/29/world/asia/29news.html*.

21. Steve Coll, "In the Gulf, Dissidence Goes Digital; Text Messaging is the New Tool of Political Underground," *Washington Post*, 29 March 2005.

22. John Kelly and Bruce Etling, "Mapping Iran's Online Public: Politics and Culture in the Persian Blogosphere," Berkman Center for Internet and Society, April 2008; available at *http://cyber.law.harvard.edu/sites/cyber.law.harvard.edu/files/Kelly&Etling_Mapping_Irans_Online_Public_2008.pdf*.

23. Omid Habibinia, "Who's Afraid of Facebook?" 3 September 2009; available at *http://riseoftheiranianpeople.com/2009/09/03/who-is-afraid-of-facebook*.

24. In addition to Freedom House, *Freedom on the Net*, see Ronald Deibert, John Palfrey, Rafal Rohozinski, and Jonathan Zitrain, *Access Denied: The Practice and Policy of Global Internet Filtering* (Cambridge: MIT Press, 2008). For the ongoing excellent work of the OpenNet Initiative, see *http://opennet.net*.

25. University of Toronto Citizen Lab, "Everyone's Guide to By-Passing Internet Censorship," September 2007; available at *www.civisec.org/guides/everyones-guides*.

26. Center for International Media Assistance, National Endowment for Democracy, "The Role of New Media in the 2009 Iranian Elections," July 2009, 2; available at *http://cima.ned.org/wp-content/uploads/2009/07/cima-role_of_new_media_in_iranian_elections-workshop_report.pdf*.

27. Rory Cellan-Jones, "Hi-Tech Helps Iranian Monitoring," BBC News, 22 June 2009. Available at *news.bbc.co.uk/2/hi/technology/8112550.stm*.

28. Rebecca MacKinnon, "Are China's Demands for Internet 'Self-Discipline' Spreading to the West?" McClatchy News Service, 18 January 2010; available at *www.mcclatchy-dc.com/2010/01/18/82469/commentary-are-chinas-demands.html*.

29. Daniel Calingaert, "Making the Web Safe for Democracy," *ForeignPolicy.com*, 19 January 2010; available at *www.foreignpolicy.com/articles/2010/01/19/making_the_web_safe_for_democracy*.

2

LIBERATION VS. CONTROL: THE FUTURE OF CYBERSPACE

Ronald Deibert and Rafal Rohozinski

Ronald Deibert *is professor of political science and director of the Canada Centre for Global Security Studies and the Citizen Lab at the Munk School of Global Affairs, University of Toronto.* **Rafal Rohozinski** *is a principal with the SecDev Group and former director of the Advanced Network Research Group of the Cambridge Security Programme. They are cofounders of the OpenNet Initiative and the Information Warfare Monitor. Their forthcoming book is* Ghost in the Machine: The Battle for the Future of Cyberspace. *The following essay is adapted from their "Cyber Wars" in the* Index on Censorship *(2010), available at* www.indexoncensorship.org. *This version originally appeared in the October 2010 issue of the* Journal of Democracy.

Every day there seems to be a new example of the ways in which human ingenuity combines with technology to further social change. For the Green Movement in Iran, it was Twitter; for the Saffron Revolution in Burma, it was YouTube; for the "color revolutions" of the former Soviet Union, it was mobile phones. No matter how restrictive the regulations or how severe the repercussions, communities around the world have exhibited enormous creativity in sidestepping constraints on technology in order to exercise their freedoms.

Looking at the seemingly endless examples of social innovation, one might easily assume that cybertechnologies possess a special power, that they are "technologies of liberation."[1] No other mode of communication in human history has facilitated the democratization of communication to the same degree. No other technology in history has grown with such speed and spread so far geographically in such a short period of time. Twitter, to take just the latest cyberapplication as an example, has grown from an average of 500,000 tweets a quarter in 2007 to more than four-billion tweets in the first quarter alone of 2010. The continual innovations in electronic communications have had unprecedented and far-reaching effects.

Yet some observers have noted that the very same technologies which give voice to democratic activists living under authoritarian rule can also be harnessed by their oppressors.[2] Cybercommunication has made possible some very extensive and efficient forms of social control. Even in democratic countries, surveillance systems penetrate every aspect of life, as people implicitly (and perhaps unwittingly) consent to the greatest invasion of personal privacy in history. Digital information can be easily tracked and traced, and then tied to specific individuals who themselves can be mapped in space and time with a degree of sophistication that would make the greatest tyrants of days past envious. So, are these technologies of freedom or are they technologies of control?

This dichotomy is itself misleading, however, as it suggests a clear-cut opposition between the forces of light and the forces of darkness. In fact, the picture is far more nuanced and must be qualified in several ways. Communications technologies are neither empty vessels to be filled with products of human intent nor forces unto themselves, imbued with some kind of irresistible agency. They are complicated and continuously evolving manifestations of social forces at a particular time and place. Once created, technologies in turn shape and limit the prospects for human communication and interaction in a constantly iterative manner. Complicating matters further is the inescapable presence of contingency. Technical innovations may be designed for specific purposes but often end up having wildly different social uses and effects than those intended by their creators. Yet these "alternative rationalities"— systems of use based on local culture and norms, particularly those that originate outside the developed world—often become the prevailing paradigm around which technologies evolve, until they in turn are disrupted by unanticipated uses or new innovations.[3]

The concepts of "liberation" and "control" also require qualification. Both are socially constructed ideas whose meaning and thus application can vary widely depending on the context in which they appear. Different communities work to be free (or "liberated") from different things—for example, colonial rule or gender or religious discrimination. Likewise, social control can take many forms, and these will depend both on the values driving them as well as what are perceived to be the objects of control. Countless liberation movements and mechanisms of social control coexist within a shared but constantly evolving communications space at any one time. This makes any portrayal of technology that highlights a single overarching characteristic biased toward either liberation or control seem fanciful.

This social complexity is a universal characteristic of all technological systems, but it is especially marked in the communications arena for several reasons. Processes of globalization, which are both products of and contributors to cyberspace, intensify the mix of actors, cultures, interests, and ideas in the increasingly dense pool of communications. Although it may seem clichéd to note that events on one side of the planet can ripple back at the speed of light to affect what happens on the other side, we must not underestimate the proliferation of players whose actions help to shape cyberspace and who in turn are shaped by their own interactions within cyberspace. This "dynamic density" also accelerates the pace of change inherent in cyberspace, making it a moving target.[4] Innovations, which potentially may come from any of the millions of actors in cyberspace, can occur daily. This means that rather than being a static artifact, cyberspace is better conceptualized as a constantly evolving domain—a multilevel ecosystem of physical infrastructure, software, regulations, and ideas.

The social complexity of cyberspace is compounded by the fact that much of it is owned and operated by thousands of private actors, and some of their operations cross national jurisdictions. Guided by commercial principles, these enterprises often make decisions that end up having significant political consequences. For example, an online chat service may handle or share user data in ways that put users in jeopardy, depending on the jurisdiction in which the service is offered. Such considerations are especially relevant given the current evolution toward "cloud computing" and software-as-a-service business models. In these models, information and the software through which users interact are not physically located on their own computers but are instead hosted by private companies, often located in faraway jurisdictions. As a result, we have the curious situation in which individuals' data are ultimately governed according to laws and regulations over which they themselves have no say as citizens. This also accelerates existing trends toward the privatization of authority.[5]

Although the decisions taken by businesses—the frontline operators in cyberspace—play a critical role, cyberspace is also shaped by the actions of governments, civil society, and even individuals. Because corporations are subject to the laws of the land in which they operate, the rules and regulations imposed by national governments may inadvertently serve to carve up the global commons of information. According to the OpenNet Initiative research consortium, more than forty countries, including many democracies, now engage in Internet-content filtering.[6] The actions of civil society matter as well. Individuals, working alone or collectively through networks, can create software, tools, or forms of mobilization that have systemwide implications—not all of them necessarily benign. In fact, there is a hidden subsystem of cyberspace made up of crime and espionage.

In short, the actions of businesses, governments, civil society, criminal organizations, and millions of individuals affect and in turn are affected by the domain of cyberspace. Rather than being an ungoverned realm, cyberspace is perhaps best likened to a gangster-dominated version of New York: a tangled web of rival public and private authorities, civic associations, criminal networks, and underground economies. Such a complex network cannot be accurately described in the one-dimensional terms of "liberation" or "control" any more than the domains of land, sea, air, or space can be. Rather, it is composed of a constantly pulsing and at times erratic mix of competing forces and constraints.

Liberation: From What and for Whom?

Much of the popular reporting about cyberspace and social mobilization is biased toward liberal-democratic values. If a social movement in Africa, Burma, or Iran employs a software tool or digital technology to mobilize supporters, the stories appear throughout the global media and are championed by rights activists.[7] Not surprisingly then, these examples tend to be generalized as the norm and repeated without careful scrutiny. But social mobilization can take various forms motivated by many possible rationales, some of which may not be particularly "progressive."[8] Due to both media bias and the difficulties of conducting primary research in certain contexts, these alternative rationalities tend to be obscured from popular view by the media and underexplored by academics.[9] Yet they are no less important than their seemingly more benign counterparts, both for the innovations that they produce and the reactions that they generate.

Consider, for example, the enormous criminal underworld in cyberspace. Arguably at the cutting edge of online innovation, cybercriminals have occupied a largely hidden, parasitic ecosystem within cyberspace, attacking the insecure fissures that open up within this constantly morphing domain. Although most cybercrime takes the form of petty spam (the electronic distribution of unsolicited bulk messages), the sophistication and reach of cybercriminals today are startling. The production of "malware"—malicious software—is now estimated to exceed that of legitimate software, although no one really knows its full extent. About a million new malware samples a month are discovered by security engineers, with the rate of growth increasing at a frightening pace.

One of the more ingenious and widespread forms of cybercrime is "click fraud," whereby victims' computers are infected with malicious software and redirected to make visits to online pay-per-click ads operated by the attackers. Although each click typically generates income on the order of fractions of a penny, a "botnet" (a group of thousands of infected computers referred to as "zombies") can bring in millions of dollars for the criminals.

One such cybercriminal enterprise called Koobface (an anagram of Facebook) exploits security vulnerabilities in users' machines while also harvesting personal information from Facebook and other social-networking services. It creates thousands of malicious Facebook accounts every day, each of which is then directed toward click fraud or malicious websites that prompt the download of Trojan horses (malware downloads that appear legitimate). With the latter, Koobface can extract sensitive and confidential information such as credit-card account numbers from the infected computers of unwitting users, or deploy the computers as zombies in botnets for purposes of distributed computer-network attacks. Like the mirror universe on the television series *Star Trek,* in which parallel Captain Kirks and Spocks were identical to the originals except for their more malicious personalities, these phony accounts are virtually indistinguishable from the real ones. The Koobface enterprise demonstrates extraordinary ingenuity in social networking, but directed entirely toward fraudulent ends.

Just as software, social-networking platforms, and other digital media originally designed for consumer applications may be redeployed for political mobilization, innovations developed for cybercrime are often used for malicious political activity. Our research reveals the deeply troubling trend of cybercrime tools being employed for espionage and other political purposes.

Twice in the last two years, the Information Warfare Monitor has uncovered major global cyberespionage networks infiltrating dozens of high-level political targets, including foreign ministries, embassies, international organizations, financial institutions, and media outlets. These investigations, documented in the reports "Tracking GhostNet" and "Shadows in the Clouds," unearthed the theft of highly sensitive documents and the extensive infiltration of targets ranging from the offices of the Dalai Lama to India's National Security Council. The tools and methods used by the attackers had their origins in cybercrime and are widely available on the Internet black market.[10] Indeed, "Gh0st Rat," the main device employed by the cyberespionage network, is available for free download and has been translated into multiple languages. Moreover, although the networks examined in both studies are almost certainly committing politically motivated espionage rather than crime per se, our research suggests that the attackers were not direct agents of government but were probably part of the Chinese criminal underworld, either contracted or tolerated by Chinese officials.

Likewise, the OpenNet Initiative analyzed the cyberattacks waged against Georgian government websites during the August 2008 war with Russia over South Ossetia. The computers that were harvested together to mount distributed denial-of-service attacks were actually botnets already well known to researchers studying cybercrime and fraud, and had been used earlier to attack pornography and gambling sites for purposes of extortion.[11]

The most consistent demonstrations of digital ingenuity can be found in the dark worlds of pornography, militancy, extremism, and hate. Forced to operate in the shadows and constantly maneuvering to stay ahead of their pursuers while attempting to bring more people into their folds, these dark networks adapt and innovate far more rapidly and with greater agility than their more progressive counterparts. Al-Qaeda persists today, in part, because of the influence of jihadist websites, YouTube channels, and social-networking groups, all of which have taken the place of physical meeting spaces. Just as disparate human-rights groups identify with various umbrella causes to which they belong through their immersion in social-networking services and chat platforms, so too do jihadists and militants mobilize around a common "imagined community" that is nurtured online.

Perhaps even more challenging to the liberal-democratic vision of liberation technology is that much of what is considered criminal and antisocial behavior online increasingly originates from the young online populations in developing and postcommunist countries, many of whom live under authoritarianism and suffer from structural economic inequalities. For these young "digital natives," operating an email scam or writing code for botnets, viruses, and malware represents an opportunity for economic advancement. It is an avenue for tapping into global supply chains and breaking out of conditions of local poverty and political inequality—itself a form of liberation.

In other words, regardless of whatever specific characteristics observers attribute to certain technologies, human beings are unpredictable and innovative creatures. Just because a technology has been invented for one purpose does not mean that it will not find other uses unforeseen by its creators. This is especially true in the domains of crime, espionage, and civil conflict, where innovation is not encumbered by formal operating procedures or respect for the rule of law.

Enclosing the Commons: Next-Generation Controls

Arguments linking new technologies to "liberation" must also be qualified due to the ongoing development of more sophisticated cyberspace controls. Whereas it was once considered impossible for governments to control cyberspace, there are now a wide variety of technical and nontechnical means at their disposal to shape and limit the online flow of information. Like the alternative rationalities described above, these can often escape the attention of the media and other observers. But these control mechanisms are growing in scope and sophistication as part of a general paradigm shift in cyberspace governance and an escalating arms race in cyberspace.

To understand cyberspace controls, it is important first to consider a sea-change in the ways in which governments approach the domain. During the "dot-com" boom of the 1990s, governments generally took a hands-off approach to the Internet by adhering to a *laissez-faire* eco-

nomic paradigm, but a gradual shift has since occurred. While market
ideas still predominate, there has been a growing recognition of serious
risks in cyberspace.

The need to manage these risks has led to a wave of securitization
efforts that have potentially serious implications for basic freedoms.[12]
For example, certain security measures and regulations have been put
in place for purposes of copyright and intellectual-property protection.
Although introduced as safeguards, these regulations help to legitimize
government intervention in cyberspace more generally—including in
countries whose regimes may be more interested in self-preservation
than in property protections. If Canada, Germany, Ireland, or another industrialized democracy can justifiably regulate behavior in cyberspace
in conformity with its own national laws, who is to say that Belarus,
Burma, Tunisia, or Uzbekistan cannot do the same in order to protect
state security or other national values?

The securitization of cyberspace has been driven mainly by a "defen-
sive" agenda—to protect against threats to critical infrastructures and
to enable law enforcement to monitor and fight cybercrime more effectively. There are, however, those who argue that "offensive" capabilities
are equally important. In order to best defend key infrastructures, the
argument goes, governments must also understand how to wage attacks,
and that requires a formal offensive posture. Most of the world's armed
forces have established, or are in the process of establishing, cyber-
commands or cyberwarfare units. The most ambitious is the U.S. Cyber
Command, which unifies U.S. cyber-capabilities under a separate command led by General Keith Alexander of the National Security Agency. Such an institutional innovation in the armed forces of the world's
leading superpower provides a model for similar developments in other
states' armed forces, who feel the need to adapt or risk being left behind.

Not surprisingly, there have been a growing number of incidents
of computer-network attacks for political ends in recent years, including those against Burmese, Chinese, and Tibetan human-rights organizations, as well as political-opposition groups in the countries of the
former Soviet Union. It would be disingenuous to draw a direct line
between the establishment of the U.S. Cyber Command and these incidents, especially since many of these practices have been pioneered
through innovative and undeclared public-private partnerships between
intelligence services in countries such as Burma, China, and Russia and
their emergent cybercriminal underclasses. Yet it is fair to argue that the
former sets a normative standard that allows such activities to be tolerated and even encouraged. We should expect these kinds of attacks to
grow as governments explore overt and declared strategies of offensive
action in cyberspace.

Further driving the trend toward securitization is the fact that private-
sector actors, who bear the brunt (and costs) of defending cyberspace's

critical infrastructures against a growing number of daily attacks, are increasingly looking to their own governments to carry this burden as a public good. Moreover, a huge market for cybersecurity services has emerged, estimated to generate between US$40 and $60 billion annually in the United States alone. Many of the companies that now fill this space stand to gain by fanning the flames of cyberwar. A few observers have questioned the motivations driving the self-serving assessments that these companies make about the nature and severity of various threats.[13] Those criticisms are rare, however, and have done little to stem fear-mongering about cybersecurity.

This momentum toward securitization is helping to legitimize and pave the way for greater government involvement in cyberspace. Elsewhere, we have discussed "next generation" controls—interventions that go beyond mere filtering, such as those associated with the Great Firewall of China.[14] Many of these controls have little to do with technology and more to do with inculcating norms, inducing compliant behavior, and imposing rules of the road, and they stem from a multitude of motivations and concerns. Any argument for the liberating role of new technologies needs to be evaluated in the wider context of these next-generation controls.

Legal measures. At the most basic level, government interventions in cyberspace have come through the introduction of slander, libel, copyright-infringement, and other laws to restrict communications and online activities.[15] In part, the passage of such laws reflects a natural maturation process, as authorities seek to bring rules to cyberspace through regulatory oversight. Sometimes, however, it also reflects a deliberate tactic of strangulation, since threats of legal action can do more to prevent damaging information from surfacing than can passive filtering methods implemented defensively to block websites. Such laws can create a climate of fear, intimidation, and ultimately self-censorship.

Although new laws are being drafted to create a regulatory framework for cyberspace, in some cases old, obscure, or rarely enforced regulations are cited *ex post facto* to justify acts of Internet censorship, surveillance, or silencing. In Pakistan, for example, old laws concerning "blasphemy" have been used to ban access to Facebook, ostensibly because there are Facebook groups that are centered around cartoons of Muhammad.[16] Governments have also shown a willingness to invoke national-security laws to justify broad acts of censorship. In Bangladesh, for example, the government blocked access to all of YouTube because of videos clips showing Prime Minister Sheikh Hasina defending her decision to negotiate with mutinous army guards. The Bangladesh Telecommunications Commission chairman, Zia Ahmed, justified the decision by saying: "[T]he government can take any decision to stop any activity that threatens national unity and integrity."[17] In Lebanon, infrequently used defamation laws were invoked to arrest three Facebook users for posting criticisms of the Lebanese president, in spite of constitutional protections of freedom of speech.[18] In Venezuela, several people were arrested recently after posting comments on Twitter about the country's banking system. The arrests were made based on a provision in the country's banking laws that prohibits the dissemination of "false information."[19] Numerous other examples could be cited that together paint a picture of growing regulatory intervention into cyberspace by governments, shaping and controlling the domain in ways that go beyond technical blocking. Whereas at one time such regulatory interventions would have been considered exceptional and misguided, today they are increasingly becoming the norm.

Informal requests. While legal measures create the regulatory context for denial, for more immediate needs, authorities can make informal "requests" of private companies. Most often such requests come in the form of pressure on Internet service providers (ISPs) and online hosting services to remove offensive posts or information that supposedly threatens "national security" or "cultural sensitivities." Google's recent decision to reconsider its service offerings in China reflects, in part, that company's frustration with having to deal with such informal removal requests from Chinese authorities on a regular basis. Some governments have gone so far as to pressure the companies that run the infrastructure, such as ISPs and mobile phone operators, to render services inoperative in order to prevent their exploitation by activists and opposition groups.

In Iran, for example, the Internet and other telecommunications services have slowed down during public demonstrations and in some instances have been entirely inaccessible for long periods of time or in certain regions, cities, and even neighborhoods. While there is no official acknowledgement that service is being curtailed, it is noteworthy that the Iranian Revolutionary Guard owns the main ISP in Iran—the Telecommunication Company of Iran (TCI).[20] Some reports indicate that officials from the Revolutionary Guard have pressured TCI to tamper with Internet connections during the recent crises. In authoritarian countries, where the lines between public and private authorities are often blurred or organized crime and government authority mingle in a dark underworld, such informal requests and pressures can be particularly effective and nearly impossible to bring to public account.

Outsourcing. It is important to emphasize that cyberspace is owned and operated primarily by private companies. The decisions taken by those companies about content controls can be as important as those taken by governments. Private companies often are compelled in some manner to censor and surveil Internet activity in order to operate in a particular jurisdiction, as evidenced most prominently by the collusion of Google (up until January 2010), Microsoft, and Yahoo in China's Internet censorship practices. Microsoft's Bing, which tailors its search engine to serve different countries and regions and offers its services in 41 languages, has an information-filtering system at the keyword level for users in several countries. According to research by the OpenNet Initiative's Helmi Noman, users located in the Arab countries where he tested are prevented from conducting Internet searches relating to sex and other cultural norms in both Arabic and English. Microsoft's explanation as to why some search keywords return few or no results states, "Sometimes websites are deliberately excluded from the results page to remove inappropriate content as determined by local practice, law, or regulation." It is unclear, however, whether Bing's keyword filtering in the Arab world is an initiative of Microsoft or whether any or all of the Arab states have asked Microsoft to comply with local censorship practices and laws.[21]

In some of the most egregious cases, outsourced censorship and monitoring controls have taken the form either of illegal acts or of actions contrary to publicly stated operating procedures and privacy protections. This was dramatically illustrated in the case of Tom-Skype, in which the Chinese partner of Skype put in place a covert surveillance system to track and monitor prodemocracy activists who were using Skype's chat function as a form of outreach. The system was discovered only because of faulty security on the servers operated by Tom Online. In May 2009, the Chinese government introduced new laws that required personal-computer manufacturers to bundle a filtering software with all of the computers sold in the country. Although this was strongly resisted by many companies, others willingly complied. While this requirement seems to have faded over time, it is nonetheless indicative of the types of actions that governments can take to control access points to cyberspace via private companies.

Access points such as Internet cafes are becoming a favorite regulatory target for authoritarian governments. In Belarus, ISPs and Internet cafes are required by law to keep lists of all users and turn them over to state security services.[22] Many other governments have similar requirements. In light of such regulations, it is instructive to note that many private companies collect user data as a matter of course and reserve the right in their end-user license agreement to share such information with any third party of their choosing.

Presumably, there are many still undiscovered acts of collusion between companies and governments. For governments in both the developed and developing worlds, delegating censorship and surveillance to private companies keeps these controls on the "frontlines" of the networks and coopts the actors who manage the key access points and hosting platforms. If this trend continues, we can expect more censorship and surveillance responsibilities to be carried out by private companies, carrier hotels (ISP co-location centers), cloud-computing services, Internet exchanges, and telecommunications companies. Such a shift in the locus of controls raises serious issues of public accountability and transparency for citizens of all countries. It is in this context that Google's dramatic announcement to end censorship of its Chinese search engine should be considered a watershed moment. Whether other companies follow Google's lead, and how China, other countries, and the international community as a whole will respond, are critical open questions that may help to shape the public accountability of private actors in this domain.

"Just-in-time blocking." Disabling or attacking critical information assets at key moments in time—during elections or public demonstrations, for example—may be the most effective tool for influencing political outcomes in cyberspace. Today, computer-network attacks, including the use of distributed denial-of-service attacks, can be easily marshaled and targeted against key sources of information, especially in the developing world, where networks and infrastructure tend to be fragile and prone to disruption. The tools used to mount botnet attacks are now thriving like parasites in the peer-to-peer architectures of insecure servers, personal computers, and social-networking platforms. Botnets can be activated against any target by anyone willing to pay a fee. There are cruder methods of just-in-time blocking as well, such as shutting off power in the buildings where servers are located or tampering with domain-name registration so that information is not routed to its proper destination. This kind of just-in-time blocking has been empirically documented by the OpenNet Initiative in Belarus, Kyrgyzstan, and Tajikistan, as well as in numerous other countries.

The attraction of just-in-time blocking is that information is disabled only at key moments, thus avoiding charges of Internet censorship and allowing for plausible denial by the perpetrators. In regions where Internet connectivity can be spotty, just-in-time blocking can be easily passed off as just another technical glitch with the Internet. When such attacks are contracted out to criminal organizations, determining attribution of those responsible is nearly impossible.

Patriotic hacking. One unusual and important characteristic of cyberspace is that individuals can take creative actions—sometimes against perceived threats to their country's national interest—that have system-wide effects. Citizens may bristle at outside interference in their country's internal affairs or take offense at criticism directed at their governments, however illegitimate those governments may appear to outsiders. Those individuals who possess the necessary technical skills have at times taken it upon themselves to attack adversarial sources of information, often leaving provocative messages and warnings behind. Such actions make it difficult to determine the provenance of the attacks: Are they the work of the government or of citizens acting independently? Or are they perhaps some combination of the two? Muddying the waters further, some government security services informally encourage or tacitly approve of the actions of patriotic groups.

In China, for example, the Wu Mao Dang, or 50 Cent Party (so named for the amount of money its members are supposedly paid for each Internet post), patrols chatrooms and online forums, posting information favorable to the regime and chastising its critics. In Russia, it is widely believed that the security services regularly coax hacker groups to fight for the motherland in cyberspace and may "seed" instructions on prominent nationalist websites and forums for hacking attacks. In late 2009 in Iran, a shadowy group known as the Iranian Cyber Army took over Twitter and some key opposition websites, defacing the home pages with their own messages. Although no formal connection to the Iranian authorities has been established, the groups responsible for the attacks posted pro-regime messages on the hacked websites and services.

Targeted surveillance and social-malware attacks. Accessing sensitive information about adversaries is one of the most important tools for shaping political outcomes, and so it should come as no surprise that great effort has been devoted to targeted espionage. The Tom-Skype example is only one of many such next-generation methods now becoming common in the cyber-ecosystem. Infiltration of adversarial networks through targeted "social malware" (software designed to infiltrate an unsuspecting user's computer) and "drive-by" Web exploits (websites infected with viruses that target insecure browsers) is exploding throughout the dark underbelly of the Internet. Among the most prominent examples of this type of infiltration was a targeted espionage attack on Google's infrastructure, which the company made public in January 2010.

These types of attacks are facilitated by the careless practices of civil society and human-rights organizations themselves. As Nart Villeneuve and Greg Walton have shown in a recent Information Warfare Monitor report, many civil society organizations lack simple training and resources, leaving them vulnerable to even the most basic Internet attacks.[23] Moreover, because such organizations generally thrive on awareness-raising and advocacy through social networking and email lists, they often unwittingly become compromised as vectors of attacks, even by those whose motivations are not political per se. In one particularly egregious example, the advocacy group Reporters Without Borders unknowingly propagated a link to a malicious website posing as a Facebook petition to release the Tibetan activist Dhondup Wangchen. As with computer network attacks, targeted espionage and social-malware attacks are being developed not just by criminal groups and rogue actors, but also at the highest levels of government. Dennis Blair, the former U.S. director of national intelligence, recently remarked that the United States must be "aggressive" in the cyberdomain in terms of "both protecting our own secrets and stealing those of others."[24]

A Nuanced Understanding

There are several theoretical and policy implications to be drawn from the issues we raise. First, there needs to be a much more nuanced understanding of the complexity of the communications space in which we operate. We should be skeptical of one-dimensional or ahistorical depictions of technologies that paint them with a single brush. Cyberspace is a domain of intense competition, one that creates an ever-changing matrix of opportunities and constraints for social forces and ideas. These social forces and ideas, in turn, are imbued with alternative rationalities that collide with one another and affect the structure of the communications environment. Unless the characteristics of cyberspace change radically in the near future and global culture becomes monolithic, linking technological properties to a single social outcome such as liberation or control is a highly dubious exercise.

Second, we must be cautious about promoting policies that support "freedom" software or other technologies presented as magic solutions to thorny political problems. Early on, the Internet was thought to be a truly democratic arena beyond the reach of government control. Typically, the examples used to illustrate this point related to heavy-handed attempts to filter access to information, which are relatively easy to bypass. This conventional wisdom has, in turn, led to efforts on the part of governments to sponsor "firewall-busting" programs and to encourage technological "silver bullets" that will supposedly end Internet censorship once and for all. This viewpoint is simplistic, as it overlooks some of the more important and powerful next-generation controls that are being employed to shape the global commons. Liberation, freedom, and democracy are all socially contested concepts, and thus must be secured by social and political means. Although the prudent support of technological projects may be warranted in specific circumstances, they should be considered as adjuncts to comprehensive strategies rather than as solutions in and of themselves. The struggles over freedom of speech, access to information, privacy protections, and other human-rights issues that now plague cyberspace ultimately pose political problems that are grounded in deeply rooted differences. A new software application, no matter how ingenious, will not solve these problems.

Third, we need to move beyond the idea that cyberspace is not regulated or is somehow immune to regulation. Nothing could be further from the truth. If anything, cyberspace is overregulated by the multitude of actors whose decisions shape its character, often in ways that lack transparency and public accountability. The question is not *whether* to regulate cyberspace, but rather *how* to do so—within which forum, involving which actors, and according to which of many competing values. The regulation of cyberspace tends to take place in the shadows, based on decisions taken by private actors rather than as a result of public deliberation. As the trend toward the securitization and privatization of cyberspace continues, these problems are likely to become more, rather than less, acute.

Finally, for the governance of cyberspace to be effective, it must uncover what is going on "below the surface" of the Internet, largely invisible to the average user. It is there that most of the meaningful limits on action and choice now operate, and they must be unearthed if basic human rights are to be protected online. These subterranean controls have little to do with technology itself and more to do with the complex nature of the communications space in which we find ourselves as we enter the second decade of the twenty-first century. Meaningful change will not come overnight with the invention of some new technology. Instead, it will require a slow process of awareness-raising, the channeling of ingenuity into productive avenues, and the implementation of liberal-democratic restraints.

NOTES

1. Larry Diamond, "Liberation Technology," *Journal of Democracy* 21 (July 2010): 70–84.

2. Elia Zureik et al., eds., *Surveillance, Privacy, and the Globalization of Personal Information* (McQuill-Queen's University Press, 2010).

3. Our conception of "alternative rationalities" is inspired by Ulrich Beck et al., *Reflexive Modernization* (Cambridge: Polity, 1994). The concept of alternative rationalities has its origins in Max Weber's work and is further developed in critical and postmodern theories.

4. For the concept of "dynamic density," see John Gerard Ruggie, "Continuity and Transformation in the World Polity: Toward a Neorealist Synthesis," *World Politics* 35 (January 1983): 261–85.

5. A. Claire Cutler, Virginia Haufler, and Tony Porter, *Private Authority and International Affairs* (New York: SUNY Press, 1999).

6. Ronald J. Deibert et al., eds., *Access Controlled: The Shaping of Power, Rights and Rule in Cyberspace* (Cambridge: MIT Press, 2010).

7. See, for example, "Iran's Twitter Revolution," *Washington Times*, 16 June 2009; available at *www.washingtontimes.com/news/2009/jun/16/irans-twitter-revolution*.

8. Chrisanthi Avgerou, "Recognising Alternative Rationalities in the Deployment of Information Systems," *Electronic Journal of Information Systems in Developing Countries* 3 (2000); available at *www.ejisdc.org/ojs2/index.php/ejisdc/article/view/19*.

9. Rafal Rohozinski, "Bullets to Bytes: Reflections on ICTs and 'Local' Conflict," in Robert Latham, ed., *Bombs and Bandwidth: The Emerging Relationship between Information Technology and Security* (New York: New Press, 2003), 222.

10. Information Warfare Monitor and Shadowserver Foundation, *Shadows in the Cloud: Investigating Cyber Espionage 2.0*, JR03-2010, 6 April 2010; Information Warfare Monitor, *Tracking GhostNet: Investigating a Cyber Espionage Network*, JR02-2009, 29 March 2009.

11. Ronald Deibert, Rafal Rohozinski, and Masashi Crete-Nishihata, "Cyclones in Cyberspace: Information Shaping and Denial in the 2008 South Ossetia War," ms. forthcoming.

12. Ronald Deibert and Rafal Rohozinski, "Risking Security: The Policies and Paradoxes of Cyberspace Security," *International Political Sociology* 4 (March 2010): 15–32.

13. Stephen Walt, "Is the Cyber Threat Overblown?" *Foreign Policy,* 3 March 2010; available at *http://walt.foreignpolicy.com/posts/2010/03/30/is_the_cyber_threat_overblown.*

14. Deibert et al., *Access Controlled.*

15. The following section draws from an earlier article of ours: "Cyber Wars," Index on Censorship, March 2010; available at *www.indexoncensorship.org/2010/03/cyberwars-technology-deiber.*

16. See *http://en.rsf.org/pakistan-court-orders-facebook-blocked-19-05-2010,37524. html.*

17. See *www.telegraph.co.uk/news/worldnews/asia/bangladesh/4963823/YouTube-blocked-in-Bangladesh-after-guard-mutiny.html.*

18. See *www.guardian.co.uk/commentisfree/libertycentral/2010/jul/03/lebanon-facebook-president-insult.*

19. See *www.latimes.com/technology/sns-ap-lt-venezuela-twitter,0,6311483.story.*

20. "IRGC Consortium Takes Majority Equity in Iran's Telecoms," 5 October 2009, *www.zawya.com/story.cfm/sidv52n40-3NC06/IRGC%20Consortium%20Takes%20 Majority%20Equity%20In%20Iran%26rsquo%3Bs%20Telecoms.*

21. See *http://opennet.net/sex-social-mores-and-keyword-filtering-microsoft-bing-arabian-countries.*

22. See *http://technology.timesonline.co.uk/tol/news/tech_and_web/the_web/article 1391469.ece.*

23. See *www.infowar-monitor.net/2009/10/0day-civil-society-and-cyber-security.*

24. See *www.govinfosecurity.com/p_print.php?t=a&id=1786.*

3

INTERNATIONAL MECHANISMS OF CYBERSPACE CONTROLS

Ronald Deibert

Ronald Deibert *is professor of political science and director of the Canada Centre for Global Security Studies and the Citizen Lab at the University of Toronto's Munk School of Global Affairs. He is a cofounder and a principal investigator of the OpenNet Initiative and Information Warfare Monitor projects. His forthcoming book (with Rafal Rohozinski) is* Ghost in the Machine: The Battle for the Future of Cyberspace.

One of the burgeoning areas of Internet research is the study of cyberspace controls—the implementation of government-mandated or privately implemented filtering, surveillance, and other means of shaping cyberspace for strategic ends. Whereas it was once assumed that cyberspace was immune to government regulation because of its swiftly changing nature and distributed architecture, a growing body of scholarship has shown convincingly how governments can shape and constrain access to information and freedom of speech online within their jurisdictions.

Today, more than thirty countries engage in Internet filtering, not all of them authoritarian regimes.[1] Internet-surveillance policies are now widespread and bearing down on the private-sector companies that own and operate the infrastructure of cyberspace, including Internet Service Providers (ISPs). Likewise, a new generation of second- and third-order controls complements filtering and surveillance, creating a climate of self-censorship.[2] There is a very real arms race in cyberspace that threatens to subvert the Internet's core characteristics and positive network effects.

The study of cyberspace controls has tended to focus on the nation-state as the primary unit of analysis, and has examined the deepening and widening of these controls within domestic contexts. For example, the leading international research organization dedicated to studying Internet filtering—the OpenNet Initiative (ONI)—has published an annual series of country and regional reports that are based on an empirical examination of country-level controls.[3] Its reports have become touchstones for information and analysis of Internet filtering and are important empirical contributions to the study of cyberspace controls.

Largely unexamined so far, however, are the *international* dynamics by which such controls—and the resistance to them—may spread. These dynamics and mechanisms are important to consider because states do not operate in a vacuum; they are part of an international system that has important implications for what they do and how they behave. This can have both "positive" and "negative" characteristics.[4] In a positive sense, states learn from and imitate each other. They borrow and share best practices, skills, and technologies. They take cues from what like-minded states are doing and implement policies accordingly.

On the negative side, states compete against one another. Their perceptions of adversarial intentions and threats can affect the decisions that they take. This dynamic has been characterized in the international-relations literature as the logic of the "security dilemma." One can see this logic at work today in the domain of cyberspace with the development of national military capabilities to fight and win wars in cyberspace.

The international system also comprises transnational actors—namely, civil society networks and private-sector firms—that serve as conduits and propagators of ideas and policies. Civil society networks educate users within countries about best practices and networking strategies, and operate largely irrespective of national boundaries.[5] The networks that tend to get the most attention are those promoting human rights, such as access to information, freedom of speech, and privacy rights. These networks come in a variety of shapes and sizes. Some are independent and largely grassroots in origin; others have been drawn into a support structure synchronized to the foreign-policy goals of major powers such as the United States and the European Union. But very few of them, especially the more important ones, operate only in a domestic policy setting.

Private-sector actors are responsive to and seek to develop commercial opportunities across national boundaries, and they are increasingly a part of the international system's mechanisms and dynamics of cyberspace controls. Particularly relevant in this respect is the cybersecurity market, estimated at up to US$80 billion annually. Commercial providers of networking technology have a stake in the securitization of cyberspace and can inflate threats to serve their more parochial market interests.[6] Private actors also own and operate the vast majority of the infrastructure and services that we call cyberspace. For that reason alone, the decisions that they take can have major consequences for the character of cyberspace worldwide. It is not too far a stretch to argue that some companies have the equivalent of "foreign policies" for cyberspace, in some ways going beyond individual governments in terms of scope and influence.

In this chapter, I first provide a brief summary of prior research on cyberspace controls, drawing primarily from the experiences of the ONI. I then lay out a research framework for the study of international mechanisms and dynamics of cyberspace controls. The aim is not to provide an exhaustive analysis of these mechanisms and dynamics as much as it is to sketch out a conceptual and analytical framework for further research. I lay out several areas where such mechanisms and dynamics might be found and investigated further. I turn in the conclusion to a consideration of some of the reasons why further research in this area is important for the study of cyberspace.

From Access Denied to Access Controlled

Studies of cyberspace controls have developed and matured as these practices have spread worldwide. Early in the Internet's history, it was widely assumed that the Internet was difficult for governments to manage and would bring about major challenges to authoritarian forms of rule. Over time, however, these assumptions have been called into question, as governments (often in coordination with the private sector) have erected a variety of information controls. It is now fair to say that there is a growing norm worldwide for national Internet filtering, although the rationale for implementing filtering varies widely from country to country.

Some justify Internet filtering to control access to content that violates copyright, exploits children, or promotes hatred and violence. Other countries filter access to content related to minority rights, religious movements, political opposition, and human-rights groups. Levels of government transparency and accountability as well as the filtering methods themselves vary broadly across the globe. Invariably, the private-sector actors who own and operate the vast majority of cyberspace infrastructure are being compelled or coerced to implement controls on behalf of states. In short, a sea change has occurred over the last decade in terms of cyberspace controls. But how did these norms of cyberspace control spread internationally?

The most authoritative research on Internet filtering is from the ONI, which uses a combination of technical interrogation, field research, and data-analysis methods to test for filtering in more than sixty countries on an annual basis.[7] The ONI's methods were developed very much in response to the predominant question circulating at the time of its inception (2002): Could governments control access to information online within their jurisdictions? To answer this question, the ONI built a global-level, but nationally based, testing regime. Researchers from the ONI download specially designed software that connects back to databases

at the University of Toronto. The databases contain categorized lists
of URLs, domains, keywords, and Internet services that are tested on a
regular basis across each of the major ISPs of each of the countries under consideration. The categorized lists are broken down into two main
groups: 1) a global list, which is tested in all countries and is used as a
basis to make comparative judgments across countries; and 2) a "local"
or "high-impact" list that contains URLs, domains, and keywords that
are relatively unique to a particular country context and are suspected of
being targeted for filtering in that jurisdiction.

The ONI reports provide a "snapshot" of accessibility at the time
of testing from the perspective of national information environments.
Among the findings of the ONI is that Internet filtering is growing in
scope, scale, and sophistication. The latest ONI reports indicate that
more than thirty countries engage in some form of Internet filtering, a
growing number of them being democratic, industrialized countries. The
ONI has also presented evidence of the range of techniques that states
employ to filter access to information. Some of the nondemocratic regimes that engage in Internet filtering do so using commercial filtering
products developed in the United States. Others have developed more
homegrown solutions.

The ONI has also captured the range of transparency practices through
its research. Some states provide "block pages" for banned content that
explain the rationale and legal basis for the blocking; others provide
only error pages, some of which are misleading and meant to obscure
the states' intentions. The ONI has also subjected Internet services to
scrutiny, in particular comparing the results obtained from major search
engines by requests in different countries. This has helped to expose the
collusion of Internet companies with regimes that violate human rights,
while putting pressure on those companies to become more accountable.

Recently, ONI researchers have described growing trends away from
"Chinese-style" firewall-based filtering to more subtle and fluid forms
of information control.[8] The ONI describes these as "next-generation"
methods of cyberspace controls; they include pressures on the commercial sector, outsourcing controls to private actors, and more offensive
methods, such as just-in-time attacks on key information sources and
targeted malware against opposition or adversarial groups. ONI researchers have noted that these new and subtle forms of information
control challenge the ONI's core methodology and are difficult to document empirically.

International Mechanisms and Dynamics

What is missing from the ONI's research, as well as from the growing body of scholarship on cyberspace controls, is a consideration of the
international mechanisms and dynamics of such controls. The field of

international relations is premised on the notion that there are factors that affect state behavior at the international systemic level. Put simply, states are embedded in an international order that affects what they do and how they do it. Although some of this scholarship has been rightly criticized in the past for reifying the international system and ignoring domestic-level processes, it nonetheless identifies an important dimension of political behavior that needs to be considered.[9] States' policies are formed in interaction with other states in the international system. However much domestic struggles and local threats motivate what states do, their interactions with one another, their perceptions of adversarial actions and intentions, and their placement in the international order matter as well.

International institutions. The most obvious places to look for such international dynamics are the main forums of Internet governance: the International Corporation for Assigned Names and Numbers (ICANN), the International Telecommunication Union (ITU), the Internet Governance Forum (IGF), and others. These international institutions are important touchstones for the identification of the mechanisms and dynamics that interest us here.[10] They have been studied by scholars of Internet governance who have examined the stakeholders, processes, and policy outputs of these various forums in detail for many years.[11] Yet these institutions are increasingly under new pressures as governments assert themselves more forcefully in cyberspace. As a consequence, the main issues that are addressed in these forums are changing, and previously unpoliticized or mostly technical issues are becoming the objects of intense political competition. These institutions, which may have been dismissed in the past as irrelevant or overly technical, deserve renewed attention from scholars, if only because some governments are now taking them seriously as vectors of policy formation and propagation.

For example, a loose coalition of like-minded countries has begun to develop strategic engagements with international institutions such as the ITU and the IGF in ways that are quite novel. Most strikingly, Russia and the Russian-speaking countries of the former Soviet Union have adopted a wide-ranging engagement with these forums to promote policies that synchronize with national-level laws related to information security.[12] Recently, China has explicitly stated not only that states have sovereign control over national information space, but also that global cyberspace should be governed by international institutions operating under the United Nations.[13] Not surprisingly, policies reflecting these views have been vocally supported by Hamadoun Touré, the secretary-general of the ITU, who has called for a state-based cyberarms-control treaty that would imply significant renationalization of the Internet.[14] He has also been a vocal supporter of India, Indonesia, the United Arab Emirates (UAE), and other countries that have pressured companies like

Research In Motion (RIM), the Canadian maker of Blackberry products, to share encrypted data under the rubric of national-security protections.[15] Every year since 1998, Russia has put forward resolutions at the United Nations to prohibit "information aggression," which is widely interpreted to mean ideological attempts—or the use of ideas—to undermine regime stability.[16] At least 23 countries now openly support Russia's interpretation of information security.

Sometimes engagement at these forums is intended to stifle or stonewall instead of to promote certain policies. For example, Chinese delegations have been quite prominent at IGF meetings, ironically as a means to stall this forum from gaining credibility and to undermine the broadening of Internet governance to civil society and other nonstate stakeholders. At the November 2009 IGF meeting in Egypt, for example, UN security officials disrupted a book launch of the ONI's recent volume, *Access Controlled*, because the Chinese delegation objected to a poster that contained a reference to the "Great Firewall of China."[17] The propagation of norms internationally can be facilitated not only by promoting them but also by the obstruction of contrary tendencies.

What is perhaps most interesting is that the international institutions whose missions are primarily focused on the technical coordination of the Internet—ICANN, the Internet Assigned Numbers Authority (IANA), and the regional naming authorities—have become increasingly politicized and subject to securitization pressures. For example, attendees at recent meetings of regional Internet registries have noticed the presence of government military and intelligence personnel in ways that are largely unprecedented. Governments whose strategic interests are oriented around legitimization of national controls are viewing these technical forums, once generally ignored except by specialists, as important components of a broader, more comprehensive policy engagement. For example, a coalition of Russian-speaking countries, supported by China and India, has put forward a proposal through a submeeting of the ITU to give governments veto power over ICANN decisions.[18]

Generally speaking, these countries are seen as attempting to reassert the legitimacy of national sovereign control over cyberspace by promoting such a norm at international venues. Ironically, in other words, international institutions are perceived by policy makers of these countries as vehicles of nationalization.

Policy coordination through regional organizations. Although international institutions are important conduits of norm propagation and legitimization, they can also be unwieldy and diffuse. As a consequence, coalitions of like-minded states are increasingly operating through more manageable lower-level organizations, such as regional institutions. Some of these forums attract little attention and meet in relative obscurity. Thus the actions that they take rarely see the light of day and are ignored or overlooked by activists and others concerned with Internet freedom and cyberspace governance. But the participants treat them seriously and use them as vehicles of policy coordination and information-sharing.

One example is the Shanghai Cooperation Organization (SCO), a regional organization made up of China, Kazakhstan, Kyrgyzstan, Russia, Tajikistan, and Uzbekistan.[19] India, Iran, Mongolia, and Pakistan have observer status, and Belarus and Sri Lanka are considered "dialogue partners." Iran is engaged in the SCO but prevented from formally joining because of UN sanctions. It is considered an active participant in the SCO summits, however, which have been held regularly throughout the region since the early 2000s. The SCO aims to share information and coordinate policies around a broad spectrum of cultural, economic, and security concerns, among them cyberspace policies. In 2010, the SCO issued a statement on "information terrorism" that drew attention to the way in which these countries have a shared and distinct perspective on Internet-security policy. The SCO has also engaged in joint military exercises and missions, described by some observers as simulations of how to reverse "color" revolutions and popular uprisings.[20] Unfortunately, the SCO's meetings tend to be highly secretive affairs and thus not easily subject to outside scrutiny. But they are likely to become important vehicles of policy coordination, giving unity, normative coherence, and strength to the individual countries beyond the sum of their parts.

Bilateral cooperation. Norms can diffuse internationally in the most direct way by governments sharing resources and expertise with each other in bilateral relationships. There has been longstanding speculation that China and Chinese companies are selling technology to regimes that import China's filtering and surveillance system. For example, officials from China's IT ministry recently visited Sri Lanka, ostensibly to offer advice on how to filter the Internet.[21] These discussions and arrangements are rarely transparent, however, and are typically shrouded in the type of secrecy that accompanies matters of national security, law enforcement, and intelligence. They are likely to become more important vehicles for the promotion of these states' strategic interests as they seek to propagate practices internationally that are supportive of their own domestic policies.

Informal Mechanisms

Although these forums and bilateral relations are important, by no means do they exhaust the dynamics and mechanisms of cyberspace controls at play in the international system. Here it is important to underline the many different means by which norms, behaviors, and policies are propagated internationally. Although formal sites of governance such as those above are important, norms can propagate through the international system in a variety of ways. Norm diffusion is the process through which norms are socialized and shared and then become internalized, accepted, and implemented by national actors. This process is uneven and mixed, and can vary in different contexts depending on the depth to which the norm penetrates societies. Norms enter into and are accepted into national contexts depending on preexisting belief systems of a national society that support or constrain their acceptance. Norms can be propagated internationally by norm entrepreneurs (transnational actors, NGOs, and businesses acting as conveyor belts or conduits) or through imitation, learning, socialization, and competition.[22] The latter processes are often difficult to document empirically because of their epistemic or cognitive foundations, but they are important factors in explaining the spread and adoption of policies such as Internet filtering. To understand the growth of cyberspace controls over the last decade, we need to better understand the mechanisms and dynamics of this diffusion internationally.

Imitation and learning. Among international-relations theories of all stripes, there is a basic understanding that government policies are formed on the basis of dynamic relations with other states in the international system.[23] Governments are outward-looking as much as they are inward-looking. When one government sees another doing something, pressures may build to do likewise or risk being left behind. Studies of learning and imitation in the field of international relations offer a number of hypotheses and data that can be collected and imported into the study of cyberspace controls.[24] A wealth of anecdotes suggests that this is a potentially fruitful area of inquiry.

In the most elemental sense, states learn from and imitate each other's behaviors, practices, and policies. They borrow and share best practices, skills, and technologies. They take cues from what like-minded states are doing and implement policies accordingly. Fear and "self-help" are among the most important and perennial drivers of imitation and learning. States implement policies based on reactions to what other governments are doing for fear of being left behind or overtaken by adversaries. A current example of such a dynamic can be seen clearly in the rush by many countries to pressure RIM to cooperate with local law-enforcement and intelligence agencies. After the UAE went public with its concerns that RIM might have made an arrangement with the U.S. National Security Agency that the UAE wanted extended to its own security services, numerous other governments chimed in and joined the queue, including Bahrain, India, Indonesia, and Saudi Arabia.[25]

The most intense forms of imitation and learning occur around national-security issues because of the high stakes and urgency involved. For example, in reaction to revelations of Chinese-based cyberespionage against U.S. companies and government agencies, Dennis Blair, the former U.S. director of national intelligence, argued that the United States needs to be more aggressive in stealing other countries' secrets. After major compromises to the Indian national-security and defense establishment were traced back to the Chinese criminal underground, some members of the Indian government proposed legislation to give a safe haven and stamp of approval for Indian hackers to do the same.[26] India also blocked imports of Chinese telecommunications equipment, and moved swiftly to set up cyberwarfare capabilities within its armed forces.[27] In what will be no surprise to international-relations theorists, we are now entering into a classic "security-dilemma," arms-race spiral in cyberspace, as dozens of governments look to other states' actions to justify the need to set up or bolster offensive cyberwarfare capabilities. The message sent by the establishment of the U.S. Cyber Command cannot be overemphasized in this regard. Such an institutional innovation in the armed forces of the world's largest superpower sends a major signal to the international defense community.

The imitation and learning process is not uniform, but mixes with national interests and local culture to create a warp and woof.[28] Governments can look to other states in the international system to lend legitimacy to slightly modified or even altogether different policies. For example, after the United States and other industrialized countries adopted antiterror legislation, many countries of the Commonwealth of Independent States (CIS) did likewise. Their policies, however, were much more far-reaching and oriented toward the stifling of minority-independence and political-opposition movements and the shoring up of regime stability rather than to fighting international terrorism.

A similar process can be seen in the spread of cybercrime and copyright-protection legislation. Under the umbrella of an international norm intended for one purpose, states can justify policies and actions that serve more parochial aims. Russia and other authoritarian regimes have used the excuse of copyright policing to seize opposition and NGO computers—in at least once instance with the assistance of companies like Microsoft.[29] Similarly, the now widespread belief that it is legitimate to remove videos containing "offensive" information from websites can be interpreted broadly in various national contexts. Pakistani authorities have repeatedly pressured video-hosting services to remove embarrassing or politically inflammatory videos under this rubric.

Some regimes that are geographically remote appear to be learning from one another's "best practices" when it comes to dealing with cyberspace controls over opposition groups. For example, a growing list of countries have banned SMS and instant-messaging services prior to national crises or significant events such as elections or public demonstrations. Although it is possible that each of these countries is doing so in isolation, it seems more likely that they have been inspired by the actions of other countries. Cambodia,[30] Egypt,[31] India,[32] Iran,[33] Mozambique,[34] and Turkmenistan,[35] have all disabled SMS and text messaging during or leading up to recent elections, events, and public demonstrations as a way to control social mobilization.

Imitation and learning are major components of norm propagation, but they are processes that are difficult to document empirically. Unless government representatives or policy makers specifically point to an instance or act from which they are drawing inspiration, imitation and learning processes can be obscure and have to be deduced from behavior.

Commercial conduits. Norms can spread internationally via private actors, in particular companies offering a service that supports the norm. For example, a major market for cybersecurity tools and technologies has exploded in recent years, estimated at between $60 and 80 billion annually. Companies are naturally gravitating to this expanding market in response to commercial opportunities. But they can also influence the market itself by the creation of products and tools that present new opportunities for states. There are, for example, a wide range of new products that offer "deep packet inspection" and traffic-shaping capabilities, even though such activities are contrary to fading norms of "network neutrality." There are also companies that offer services and products designed for offensive cybernetwork attacks. Naturally, the principals of these companies have a vested interest in ensuring that the market continues to expand, which can, in turn, influence government policies.

The market for surveillance and offensive computer operations that has emerged in recent years was preceded by a relatively smaller market for filtering technologies. The latter were developed initially to serve business environments but quickly spread to governments looking for solutions for Internet-censorship demands. ONI research throughout the 2000s documented a growing number of authoritarian countries using U.S.-based commercial-filtering products, including Smartfilter in Iran and Tunisia, Websense in Yemen, and Fortinet in Burma. Some of these products appear to have been tailored to meet the unique requirements of authoritarian regimes. For example, the Websense product had built-in options for filtering categories that included human-rights and non-governmental organizations. In one case, a PowerPoint presentation by Cisco (the maker of telecommunications-routing equipment) surfaced which made the argument that a market opportunity had presented itself for the company to work in collusion with China's security services.[36] Commercial solutions such as these can help to structure the realm of the possible for governments. Whereas in the past it might have been difficult or even inconceivable to engage in deep packet inspection or keyword-based filtering on a national scale, commercial solutions open up opportunities for policy makers looking to deal with vexing political problems on a fine-grained scale.

International Vacuums (*horror vacui*)

One of the least obvious mechanisms of norm propagation is the absence of restraints. Policies and behaviors can spread internationally when there are no countervailing safeguards or checks. Norm diffusion through the absence of restraints might be likened to the principle of nature abhorring a vacuum. Practices and behaviors fill a void in the policy arena. This mechanism is perhaps the most difficult to pin down empirically because it lacks any identifiable source or location. Yet it may be among the most important international dynamics of cyberspace controls.

One might hypothesize that norm diffusion via the absence of restraints is most amenable to the diffusion of "bad" norms precisely because there are no countervailing restraints. For example, the spread of cybercrime and the blurring of cybercrime and espionage can be explained in part by the ways in which the perpetrators are able to exploit fissures in the international system. Bad actors act globally and hide locally in jurisdictions where state capacity is weak and they are beyond the reach of the victims' local law enforcement. Some governments, through their *inaction,* may even be deliberately cultivating a climate favorable for crime and espionage to flourish. For example, major cyberespionage networks and acts of cybercrime have been traced back to China, Russia, and other countries that take few or, at best, symbolic measures in response, in part because of the strategic benefits that accrue to these countries from the flourishing of those activities. These governments can reap windfalls from the ecology of crime and espionage through the black market while maintaining a relatively credible position of plausible deniability.[37]

Focusing on the international dynamics and mechanisms of cyberspace controls is important for several theoretical and practical reasons. First, there are unique processes that occur at the international level that are distinct from what happens domestically. These mechanisms and dynamics help to explain why a norm for Internet filtering and surveillance is spreading internationally. States do not operate in isolation; they are part of a dense network of relations that influences their decisions and actions. Without considering these mechanisms and dynamics, we may miss some of the more important explanations for growing cyberspace controls, which have until now been primarily attributed to domestic-level causes. The framework laid out above is meant to be a first step in identifying some of the most important sources of those mechanisms and dynamics.

The focus on the spread of cyberspace controls, as outlined here, may offer an important contribution to the study of international norm diffusion more generally. Up until now, scholarship in this area has been focused almost entirely on the propagation and diffusion of "good" norms, such as landmine and chemical-weapons bans, the abolition of slavery, and the spread of democratic values.[38] The examples laid out here show that the propagation and diffusion of "bad" norms can happen along the same lines, employing some of the same mechanisms and dynamics. Further research into the spread of cyberspace controls may shed light on some unique mechanisms and dynamics employed by authoritarian and competitive authoritarian regimes. It is sometimes assumed that these governments are by definition inward-looking and have an aversion to internationalism and multilateralism. Some of the examples pointed to here show that, to the contrary, these regimes have very active international and regional engagements that are likely to continue to grow.

Finally, a focus on international mechanisms and dynamics underscores the iterative and relational quality of state behavior. States' actions and behaviors are formed very much in response to other states' decisions, often in unintended ways. This observation has important policy implications for democratic industrialized countries. The policies that these governments implement may be used by authoritarian regimes to legitimize their actions at home in ways considerably different than democratic countries originally intended. Unfortunately, there is not a lot that can be done to guard against this dynamic. But it is important to be aware of it and recognize it when it occurs. General statements about the "war on terror" or "copyright controls" can be turned into excuses for a broad spectrum of nefarious actions by authoritarian regimes. These dynamics also underscore the importance of consistency, transparency, and accountability on the part of democratic regimes. For example, shortly after U.S. secretary of state Hillary Clinton admonished governments for pressuring RIM to collude with security services, the Obama administration introduced legislation that would put in place precisely the same procedures as those requested by India, Saudi Arabia, the UAE, and others. Governments are embedded in an international system and thus a dynamic network of relations. One cannot understand the spread of cyberspace controls without understanding its international mechanisms and dynamics.

NOTES

1. Ronald J. Deibert et al., eds., *Access Denied: The Practice and Policy of Global Internet Filtering* (Cambridge: MIT Press, 2008).

2. Ronald J. Deibert et al., eds., *Access Controlled: The Shaping of Power, Rights, and Rule in Cyberspace* (Cambridge: MIT Press, 2010).

3. The author is one of the founders and principal investigators of the ONI.

4. By "positive" and "negative," I do not mean to imply a normative judgment of the policies, but rather to describe the processes around which policies are formed.

5. Margaret E. Keck and Kathryn A. Sikkink, *Activists Beyond Borders: Advocacy Networks in International Politics* (Ithaca: Cornell University Press, 1999).

6. Stephen M. Walt, "Is the Cyber Threat Overblown?" ForeignPolicy.com, 30 March 2010, available at *http://walt.foreignpolicy.com/posts/2010/03/30/is_the_cyber_threat_overblown*.

7. See *http://opennet.net*.

8. Deibert et al., *Access Controlled*.

9. For criticism along these lines, see Robert O. Keohane, ed., *Neorealism and Its Critics* (New York: Columbia University Press, 1986).

10. For a discussion of how international organizations can influence state behavior in ways not intended as part of their original design, see Michael Barnett and Martha Finnemore, "The Politics, Power and Pathologies of International Organizations," *International Organization* 53 (Autumn 1999): 699–732,

11. See, in particular, Milton L. Mueller, *Networks and States: The Global Politics of Internet Governance* (Cambridge: MIT Press, 2010).

12. Policy statement made by Igor Shchegolev, Russian minister of telecommunication and mass communications, on 4 October 2010 at the International Telecommunication Union plenipotentiary conference in Guadalajara, Mexico, available at *www.itu.int/plenipotentiary/2010/statements/russian_federation/shchegolev-ru.html*.

13. See Brenden Kuerbis, "Reading Tea Leaves: China Statements on Internet Policy," Internet Governance Project, 8 June 2010, available at *http://blog.internetgovernance.org/blog/_archives/2010/ 6/8/4548091.html*.

14. Tim Gray, "U.N. Telecom Boss Warns of Pending Cyberwar," *TechNewsDaily*, 10 September 2010, available at *www.msnbc.msn.com/id/39102447/ns/technology_and_science-security*.

15. "RIM Should Open Up User Data: UN Agency," CBC News, 2 September 2010, available at *www.cbc.ca/money/story/2010/09/02/rim-user-data-un.html*.

16. Tom Gjelten, "Seeing the Internet as an 'Information Weapon,'" NPR, 23 September 2010, available at *www.npr.org/templates/story/story.php?storyId=130052701&sc=tw&cc=share*.

17. Jonathan Fildes, "UN Slated for Stifling Net Debate," BBC News, 16 November 2009, available at *http://news.bbc.co.uk/2/hi/technology/8361849.stm, accessed 6 Oct. 2010*.

18. Gregory Francis, "Plutocrats and the Internet," CircleID, 4 October 2010, available at *www.circleid.com/posts/20101004_plutocrats_and_the_internet*.

19. See Andrew Scheineson, "The Shanghai Cooperation Organization," Council on Foreign Relations, 24 March 2009, available at *www.cfr.org/publication/10883/shanghai_cooperation_organization.html*.

20. Richard Weitz, "What's Happened to the SCO?" *The Diplomat*, 17 May 2010, available at *http://the-diplomat.com/2010/05/17/what's-happened-to-the-sco/*.

21. Bandula Sirimanna, "Chinese Here for Cyber Censorship," *Sunday Times*, 14 February 2010, *www.sundaytimes.lk/100214/News/nws_02.html*.

22. Martha Finnemore and Kathryn Sikkink, "International Norm Dynamics and Political Change," *International Organization* 52 (Autumn 1998): 887–917.

23. Kenneth N. Waltz, *Theory of International Politics* (Reading, Mass.: Addison-Wesley, 1979); and Keohane, *Neorealism and Its Critics*.

24. Benjamin E. Goldsmith, "Imitation in International Relations: Analogies, Vicarious Learning, and Foreign Policy," *International Interactions* 29, no. 3 (2003): 237–67.

25. "Factbox: BlackBerry Under Fire from States Seeking Access," Reuters, 13 August 2010, *www.reuters.com/article/2010/08/13/us-blackberry-access-factbox-idUS-TRE67B22T20100813*.

26. Harsimran Singh and Joji Thomas Philip, "Spy Game: India Readies Cyber Army to Hack into Hostile Nations' Computer Systems," *Economic Times,* 6 August 2010, available at *http://articles.economictimes.indiatimes.com/2010-08-06/news/27590170_1_computer-systems-spy-game-hackers*.

27. Rhys Blakely, "India Blocks Deals with Chinese Telecoms Companies over Cyber-Spy Fears," *Times Online,* 10 May 2010, available at *http://citizenlab.org/2010/05/india-blocks-deals-with-chinese-telecoms-companies-over-cyber-spy-fears/*.

28. For a general discussion, see Jeffrey Legro, "Which Norms Matter? Revisiting the 'Failure' of Internationalism," *International Organization* 51 (Winter 1997): 31–63.

29. Clifford J. Levy, "Russia Uses Microsoft to Suppress Dissent," *New York Times*, 22 September 2010.

30. Prerna Mankad, "Cambodia Bans Text Messaging," ForeignPolicy.com, 30 March 2007, *http://blog.foreignpolicy.com/posts/2007/03/30/cambodia_bans_text_messaging*.

31. See *www.almasryalyoum.com/en/news/restrictions-placed-sms-messages-avert-promoting-anti-regime-incitations*.

32. Harmeet Shah Singh, "India's Top Court Delays Decision on Holy Site," CNN, 23 September 2010, available at *http://edition.cnn.com/2010/WORLD/asiapcf/09/23/india.holy.verdict/index.html*.

33. Nazila Fathi, "Iran Disrupts Internet Service Ahead of Protests," *New York Times,* 11 February 2010, available at *www.nytimes.com/2010/02/11/world/middleeast/11tehran.html?ref=global-home*.

34. Janet Gunter, "Mozambique: Government Interference in SMS Service," *Global Voices,* 21 September 2010, available at *http://advocacy.globalvoicesonline.org/2010/09/21/mozambique-government-interference-in-sms-service/*.

35. Annasoltan, "Technology and Tradition Are Not Enemies: Agent.mail.ru Banned in Turkmenistan!" 1 September 2010, available at *http://www.neweurasia.net/media-and-internet/technology-and-tradition-are-not-enemies-agentmailru-banned-in-turkmenistan*.

36. Brad Reese, "Powerpoint Presentation Appears to Implicate Cisco in China," *NetworkWorld*, available at *www.networkworld.com/community/node/27957*.

37. See Information Warfare Monitor and Shadowserver Foundation, "Shadows in the Cloud: Investigating Cyber Espionage 2.0," joint report, 6 April 2010.

38. One exception is Ryder McKeown, "Norm Regress: US Revisionism and the Slow Death of the Torture Norm." *International Relations* 23 (March 2009): 5–25.

4

WHITHER INTERNET CONTROL?

Evgeny Morozov

Evgeny Morozov *is a Bernard L. Schwartz Senior Fellow at the New America Foundation and a contributing editor of* Foreign Policy, *where he runs the "Net Effect" blog* (http://neteffect.foreignpolicy. com). *He is the author of* The Net Delusion: The Dark Side of Internet Freedom *(2011). This essay originally appeared in the April 2011 issue of the* Journal of Democracy.

Leading liberal democracies such as the United States have begun promoting "Internet freedom" and, by extension, opposing "Internet control." But what exactly is this control, and how best may it be combated? As a category, it is broad, encompassing both censorship (which violates the right to free expression) and surveillance (which violates the right to privacy). This dual character of control explains why it is often so hard to assess innovations such as social networking in the abstract: They work in favor of freedom of expression by making it easier for us to express ourselves, but at the same time they also tend to work in favor of surveillance by making more of our private information public.

In addition to its ability to manifest itself as both censorship and surveillance, "Internet control" has a technological dimension and a sociopolitical dimension that often overlap in practice even though they are analytically distinct from each other. A good example of the technological control would be a national-level scheme in which government-sanctioned Internet filters automatically banned access to all sites whose URLs contained certain sensitive keywords. A good example of sociopolitical control would be a law that treated blogging platforms such as WordPress or LiveJournal as mass-media organs and made them screen all user-submitted online content prior to publication. In the former case, a government would be using technology to chill the freedom of expression directly; in the latter case, the sought-after effect would be the same, but would be indirect and mediated through the power of law rather than sought through the direct application of technology alone.

Most talk of "liberation technologies" as ways of weakening "Internet control" turns out to be about the technological rather than the sociopolitical dimension. But what if success in that area is met with larger and more sophisticated efforts at exerting sociopolitical control? Scholars still know little about the factors that influence the dynamics and the distribution of the two kinds of control. As technological methods lose efficacy, sociopolitical methods could simply overtake them: An authoritarian government might find it harder to censor blogs, but still rather easy to jail bloggers. Indeed, if censorship becomes infeasible, imprisonment may become inevitable.

Thus, if the technological dimension of Internet control were one day to be totally eliminated, the upshot could be a set of social and political barriers to freedom of expression that might on balance be worse—not least because "liberation technologies" would be powerless to overcome them. It would be a cruelly paradoxical outcome indeed should liberation technology's very success spur the creation of a sociopolitical environment in which there would be nothing for technology to "liberate."

But suppose that we could set such concerns aside. What are the ways to minimize the technological dimension of Internet censorship? On first sight, this looks like a mere engineering challenge. It may even be tempting to reframe this question as follows: Given what we know about how the Internet works, what can we do to help bypass such technological restrictions as authoritarian governments might put in place?

The proliferation of numerous censorship-circumvention technologies over the last decade suggests that even the most sophisticated Internet-filtering schemes are not immune to the ingenuity of the engineering community. The porousness and decentralization that are basic to the Internet's design make it hard to come up with a firewall that works 100 percent of the time. Unless they are forced to deal with a national Intranet featuring a fixed number of government-run websites, those with the requisite know-how will always be able to circumvent government bans by connecting to third-party computers abroad and using them to browse the uncensored Internet.

It might seem, then, that the only outstanding problems are technological in nature: making sure that tools deliver on their promises—including promises to keep their users safely undetected and anonymous—and that they remain cheap and easy to use. It might also seem that money and engineering talent would be all that is needed to solve such problems. For example, Shiyu Zhou, the founder of a Falun Gong technology group that designs and distributes software for accessing sites banned by the Chinese government, says that "the entire battle over the Internet has boiled down to a battle over resources."[1]

This is a misleading view. The sociopolitical environment will always affect the scope and intensity of technology-based efforts to get around Internet controls. Some of the constraints on the use and proliferation of such tools are anything but technological in origin, and are not traceable to resource scarcity. A tool that can help dissidents in authoritarian states to access websites that authoritarian governments have banned may also allow terrorists or pedophiles to access online resources that democratic governments have placed off-limits in keeping with their own laws and systems of due process. Similarly, any tool that allows dissidents to hide their digital doings from the prying eyes of an unfree regime's secret police may also be used by criminals to evade the legal monitoring efforts of legitimate law-enforcement agencies in liberal-democratic states. Technology and engineering, in other words, do not operate in a vacuum. The social and political environment will inevitably have much to do with determining how, where, how quickly and widely, and to what ends they are brought to bear, as well as what the public thinks about them and their uses.

Considering that many such tools are developed by activists (often working as volunteers) who have a stake in many different projects—not all of them viewed altogether favorably by governments or publics—it is unsurprising that the going can be tough. When we consider that at least one of the key people doing work on Tor—a much-celebrated system of servers and software that is designed to ensure users' online anonymity and that enjoys U.S.-government funding—has also collaborated with WikiLeaks,[2] we are not shocked to learn that the U.S. government may be having some second thoughts about this particular surveillance-evading tool. Although this has not so far had a tangible impact on the level of support that the U.S. government has been giving to Tor, that may change in the future—especially as competing projects acquire powerful backers and lobbyists eager to defend their cause in Washington.

At a minimum, any policy initiative that aims to address the technological dimension of "Internet control" needs to find a way to model the sociopolitical environment—including the tricky human element—in which such tools are designed and distributed. At this point, it is hard to predict how Western governments will react if Tor solves its functionality problems and suddenly becomes more user-friendly and faster. Nor do we know how upcoming legislation aimed at forcing Internet companies to create "backdoors" through which U.S. law-enforcement and intelligence agencies can secretly access online services such as Skype and Gmail may impede the wider deployment of a tool such as Tor, which would probably help to keep these "backdoors" closed.[3] This is not only, or even mainly, a battle for resources; there are many unresolved political issues involved here. The U.S. State Department, as Hillary Clinton announced in a widely publicized January 2010 speech,[4] may back an Internet-freedom agenda, but can the same be said for all arms of the U.S. government?

One possible solution would be to design specialized tools that would tackle Internet censorship only in particular countries such as China, Iran, or Kazakhstan. Such tools would not abet the terrorists and criminals that worry U.S. authorities—unless, of course, the bad actors managed to set up shop in one or more of those countries. Another disadvantage is that any tool with a particular geographic focus will end up becoming far more politicized than would any generic solution such as Tor.

The focus on abetting censorship circumvention in a particular country may only result in the government aiming sharper scrutiny at such tools and those who wield them. A case in point is Haystack, an anticensorship tool that U.S. "hacktivists" devised for Iranian dissidents to use in the wake of their country's 2009 "Green Wave" postelection protests. Since Haystack's users were presumed to be dissidents downloading Human Rights Watch reports rather than illegal online pornography (a common use of general-purpose tools such as Tor), the Iranian government had a particularly strong incentive to monitor them. (Haystack shut down in September 2010 after proving to be less reliable than its inventors had claimed.)

The further segmentation of this market, with the appearance of tools specific to Fiji and Tajikistan, for instance, would also make it hard to vouch for the security of each tool. Ideally, any effort to create a new country-specific tool would conform to appropriate standards and procedures, so that its technical merits could be independently assessed by third parties and subjected to peer review. The Haystack debacle suggests the pitfalls that lie in wait if such rigorous protocols are ignored.

The Sociopolitical Dimension

Internet-filtering is just one of the many options available to governments. It is also the one that is easiest to document. Moreover, it lends itself nicely to straightforward assessments by Freedom House, the OpenNet Initiative, and Herdict Web, a Harvard-based initiative that seeks input from Internet users worldwide in order to "crowdsource" real-time data regarding Internet control. Because of this relative transparency, and because being known for Web-filtering looks bad (few countries want to be spoken of in the same sentence with phrases such as the "Great Firewall of China"), governments are now experimenting with more sophisticated ways of exerting control that are harder to detect and document. These include:

Distributed denial-of-service attacks. Although reliable statistical data is scarce, anecdotal evidence suggests that politically motivated "distributed denial-of-service" (DDoS) attacks are on the rise.[5] These target individuals or entire organizations by flooding their websites with crippling volumes of artificially generated Internet traffic. This effectively shuts down the targeted site for a time and denies access to legitimate users.

The publishers of a site that comes under DDoS attack must not only scramble for ways to keep content available (especially if the assault comes at a sensitive moment such as during a period of postelection protests), but must also cope with the anger of the Internet-hosting companies that are often the ones left dealing with the consequences of such attacks. Repeated DDoS attacks on a site may eventually make it "unhostable"—no hosting company will touch it for fear of the costly cyberassaults that it will draw.

For content producers, DDoS attacks could have far worse consequences than would attempts to filter their websites in a given jurisdiction. First, a successful DDoS attack makes content unavailable anywhere and everywhere, not just in this or that place with a Web-filtering system in place. Second, DDoS assaults put heavy psychological pressure on content producers, suddenly forcing them to worry about all sorts of institutional issues such as the future of their relationship with their Internet-hosting company, the debilitating effect that the unavailability of the site may have on its online community, and the like.

From the perspective of those ordering the attacks—it is a fair assumption that in some cases this means authoritarian governments—DDoS assaults beat censorship by virtue of being far harder to trace to their source. Indeed, it can sometimes be hard to say if a DDoS attack has taken place at all. Websites go down all the time for a variety of reasons: legitimate spikes in traffic, server failures, power outages, and numerous other causes that have nothing to do with cyberattacks. Among other things, this means that there are still no reliable ways to gauge the frequency and intensity of DDoS attacks. They might be a worse threat to global freedom of expression than we currently realize.

Unfortunately, things will probably get worse, as there has already emerged a black market in DDoS attacks (you can go to eBay to "rent" the ability to launch one). This seems to be happening mainly because DDoS attacks can also be used to target businesses for the purposes of cyberextortion. Yet a recent survey by the computer-security firm Symantec indicates that 53 percent of critical-infrastructure providers report experiencing what they perceive as politically motivated cyberattacks against their networks. Those who reported having been attacked further reported an average of ten incidents over the preceding five years, with an average total cost across all five years of US$850,000.[6]

Deliberate erosion of online communities' social capital. Among the things about the Internet that make authoritarian rulers uneasy are its powers to boost civic associations of all kinds and to connect previously unconnected people and organizations with one another. Governments fearful of Internet-enabled "connectivity" have learned that censorship is too blunt an instrument to be their best weapon against communities that begin online and remain outside the state's control. Censorship can be too easily circumvented and it can backfire if the presence of a threat from the state ends up strengthening rather than weakening a target community's internal bonds. There are numerous ways to weaken community ties more effectively.

One option may be the launching of DDoS attacks as a means of shutting sites down periodically and, worse, forcing them to find a way to pay for better hosting services. Even simpler methods may include trolling or dispatching new members to create artificial splits within the community as well as intentionally provoking community administrators to take harsh and unpopular measures. These last two stratagems are labor-intensive and costly ways to hamstring and manipulate online communities, but they are more likely to prove effective than censorship.

The Chinese government has become notorious for its "fifty-centers"—people who are paid piece rates to post progovernment comments on message boards and other widely read online forums. Their work is plainly meant to influence the intellectual dynamics of online communities and sow doubts within their ranks. Vladimir Putin's Russia, likewise, has plenty of Kremlin-friendly youth movements whose members will defend the government and its policies online, including on the websites of critics.

The "nationalization" of cyberspace. In the months since Hillary Clinton's speech on Internet freedom, many governments seem to have woken up to the possibility that the United States might be keen to exploit its existing dominance of cyberspace in order to promote a certain political agenda. Whether or not their concerns are justified, the governments of China, Iran, Russia, and many other countries have suddenly realized the degree to which their own citizens are dependent on Internet services offered by U.S. companies. Editorials in their state-owned newspapers increasingly speak of "informational sovereignty," by which they mean the ability of their digital economies to function independent of foreign service providers.

In keeping with this, these governments have begun bolstering their domestic Internet enterprises at the expense of foreign competitors. Turkey made the first move into this space with its late-2009 launch of the Anabena project, which is meant to create a national search engine that better caters to "Turkish sensibilities," with a national e-mail system to follow. Iran quickly followed suit, banning Gmail in February 2010 and announcing its own national e-mail system. Later in the same year, Russia announced plans similar to Turkey's, including the establishment of a national e-mail service and the allocation of $100 million to explore the feasibility of a national search engine.

It would not be surprising to see the Chinese, Russian, and other governments declare that Internet-search services are a "strategic industry" like energy and transport and move to block foreign companies in this area. If the impression that Twitter and Facebook can facilitate political revolutions continues to gain currency, social-networking and microblogging services may end up in the "strategic" category as well. This will almost certainly be bad news for users, since local alternatives to Google, Facebook, and Twitter are likely to have more restrictive attitudes toward freedom of expression and privacy. Even if we see no new national search engines, existing local competitors to Google (China's Baidu and Russia's Yandex, for instance) may grow stronger as a result.

The outsourcing of Internet control to third parties. One way for governments to avoid direct blame for exercising more Internet control is to delegate the task to intermediaries. At a minimum, this will involve making Internet companies that offer social-networking sites, blogging platforms, or search engines take on a larger self-policing role by holding them accountable for any content that their users post or (in the case of search engines) index and make available.

Being able to force companies to police the Web according to state-dictated guidelines is a dream come true for any government. The companies must bear all the costs, do all the dirty work, and absorb the users' ire. Companies also are more likely to catch unruly content, as they are more decentralized and know their own online communities better than do the state's censors.

It would be a mistake to think that only authoritarian governments harbor such ambitions. The Italian government has been holding YouTube accountable for the videos that are published on its site. This creates an enabling environment in which authoritarian governments can justify their actions by referring to similar developments in democratic societies.

Other ways in which third parties abet Internet control are appearing as well. Thailand's strict laws against *lèse majesté* ban the publication of anything (even a Twitter post) that may offend the country's royal family. When the Thai blogosphere's rapid expansion began outstripping the authorities' monitoring capacities, a member of parliament proposed a solution in early 2009. A site called ProtectTheKing.net was set up so that Thai users could submit links to any website that they deemed offensive to the monarchy. According to the BBC, the government blocked five-thousand submitted links in the first twenty-four hours. Not surprisingly, the site's creators "forgot" to provide a way in which to complain about sites that were blocked in error.[7]

Similarly, Saudi Arabia allows its citizens to report to the Communications and Information Technology Commission any links that they find offensive; citizens do so at an average rate of about 1,200 times per day. This allows the Saudi government to achieve a certain efficiency in the censorship process. According to *Business Week,* in 2008 the Commission's censorship wing employed only 25 people, although many of them were graduates of top Western universities such as Harvard and Carnegie Mellon.[8] But many similar efforts are also emerging and flourishing organically, without any apparent state involvement. Thus, a well-coordinated group of two-hundred culturally conservative volunteers that calls itself "Saudi Flagger" regularly monitors all Saudi Arabia–related videos uploaded to YouTube. Their practice is to complain *en masse* about any videos that they do not like—mostly these contain criticisms of Islam or Saudi rulers—by "flagging" them for YouTube's administrators as inappropriate and misleading.[9] A member, Mazen Ali Ali, described this in 2009 as "perform[ing] our duty towards our religion and homeland."[10]

Private-sector innovations. The Internet-control activities of secret-police officials in authoritarian states are increasingly getting a boost from breakthroughs in data analysis that the Web itself is making cheaper to obtain. It is not only text-messaging traffic that is becoming easier to search, organize, and act on: Video footage is moving in that direction as well, thus paving the way for even more video surveillance. This explains why the Chinese government keeps installing video cameras in its most troubling cities. Not only do such cameras remind passers-by about the panopticon that they inhabit, they also supply the secret police with useful clues. In mid-2010, there were 47,000 cameras scanning Urumqi, the capital of China's restive Xinjiang Province, and that number was projected to rise to 60,000 by the end of the year.[11]

Such an expansion of video surveillance could not have happened without the involvement of Western partners. Researchers at UCLA, funded in part by the Chinese government, have managed to build surveillance software that can automatically annotate and comment on what it sees, generating text files that can later be searched by humans, obviating the need to watch hours of video footage in search of one particular frame.[12] (To make that possible, the researchers had to recruit twenty graduates of local art colleges in China to annotate and classify a library of more than two-million images.) Such automated systems are crucial in order for video surveillance to be massively "scaled up" in a useful way, since it makes sense to add new cameras only if their footage can be rapidly indexed and searched.

The maddening pace of innovation in data analysis is poised to make surveillance far more sophisticated, with new features that seem straight out of science fiction. Digital surveillance will receive a significant boost as face-recognition techniques improve and enter the consumer market. The trade in face-recognition technologies is so lucrative that even giants such as Google cannot resist getting into the game as they feel the growing pressure from smaller players such as Face.com, a popular tool that allows users to find and automatically annotate unique faces as these appear throughout their photo collections. In 2009, Face.com launched a Facebook application that first asks users to identify a Facebook friend in a photo and then proceeds to search the entire social-networking site for other pictures in which that friend appears. By early 2010, the company was boasting of having scanned nine billion pictures and identified 52 million individuals.

Applications go far beyond finding photos of one's friends on Facebook. Imagine advanced face-recognition technology in the hands of the Iranian Revolutionary Guards as they seek to ferret out the identities of people photographed during Tehran street protests. That said, governments had been using face-recognition technologies (the legitimate law-enforcement applications are obvious) for some time before these tools became commercially viable. What is most likely to happen in the case of Iran is that widely accessible face-recognition technologies will empower various solo agents, cybervigilantes who may not be on the payroll of the Islamic Republic, but who would like to help its cause. Just as Thai royalists surf the Web in search of sites criticizing the monarchy or progovernment Chinese go on the lookout for problematic blog posts, so we can predict that Islamist hard-liners in Iran will be checking photos of antigovernment protests against those in the massive commercial photo banks, populated by photos and names harvested from social-networking sites, that are sure to pop up, not always legally, once face-recognition technology goes fully mainstream. The cybervigilantes may then continue stalking the dissidents, launch DDoS attacks against their blogs, or simply report them to authorities.

Search engines capable of finding photos that contain a given face anywhere on the Internet are not far off. For example, SAPIR, an ambitious project funded by the European Union, seeks to create an audiovisual search engine that would automatically analyze a photo, video, or sound recording; extract certain features to identify it; and use these unique identifiers to search for similar content on the Web. An antigovernment chant recorded on the streets of Tehran may soon be broken down into individual voices, which in turn can then be compared to a universe of all possible voices that exist on amateur videos posted on YouTube.

Or consider Recognizr, the cutting-edge smartphone application developed by two Swedish software firms that allows anyone to point their mobile phone at a stranger and immediately query the Internet about what is known about that person (or, to be more exact, about that person's face). Its developers are the first to point to the tremendous privacy implications of their invention, promising that strict controls would eventually be built into the system.[13] Nevertheless, it is hard to believe that once the innovation genie is out of the bottle, no similar rogue applications would be available for purchase and download elsewhere.

The rise of online "publicness." If there is a clear theme to much of the Internet innovation of the last decade, it is that being open to sharing one's personal information can carry big benefits. More and more of our Internet experience is customized: Google arranges our search results in part based on what we have searched for in the past, while our Facebook identity can now "travel" with us to different sites (for example, those who visit music-streaming sites such as Pandora while logged into Facebook will be able to see what music their Facebook friends like and recommend).

When Jeff Jarvis, a professor of new media at the City University of New York and a leading Internet pundit, points out the benefits of publicness, he is right: There are, indeed, tremendous advantages to sharing our location, favorite music, or reading lists with the rest of the world.

The problem is that a world where such publicness can be turned against us is not so hard to imagine—and Internet pundits are usually the last to point out that all the digital advantages come at a price. Just as Amazon recommends books to us based on the books that we have already purchased, it is not hard to think of a censorship system that makes decisions based on the pages that we have visited and the kinds of people whom we list as our friends on social-networking sites. Might it be possible that in the not-so-distant future, a banker who peruses nothing online but Bloomberg News and the *Financial Times,* and who has only other bankers as her online friends, will be left alone to do anything she wants, even browse Wikipedia pages about human-rights violations? In contrast, a person of unknown occupation, who occasion-ally reads the *Financial Times* but who is also linked to five well-known political activists through Facebook and who has written blog comments containing words such as "democracy" and "freedom," will only be al-lowed to visit government-run websites (or, if he is an important intel-ligence target, he will be allowed to visit other sites, with his online activities closely monitored).

If online advertising is anything to judge by, such behavioral preci-sion is not far away. Google already bases the ads that it shows us on our searches and the text of our e-mails; Facebook aspires to makes its ads much more fine-grained, taking into account what kind of content that we have previously "liked" on other sites and what our friends are "lik-ing" and buying online. Imagine censorship systems that are as detailed and as fine-tuned to their "users" (targets) as the behavioral advertising that we now see every day. The only difference between the two is that one system learns everything about us in order to show us more relevant advertisements, while the other one learns everything about us in order to ban us from accessing relevant pages.

By paying so much attention to the most conventional and blandest of Internet-control methods (blocking access to particular URLs), we risk missing more basic shifts in the field. Internet censorship is poised to grow in depth, looking ever more thoroughly into what we do online and even offline. It will also grow in breadth, incorporating more and more information indicators before the "censor or do not censor" decision is made. Arguably, Green Dam Youth Escort—the Chinese software that made a lot of noise in mid-2009—was a poor implementation of an extremely powerful and dangerous concept: Green Dam analyzed the kinds of activities that the user was engaged in and made a decision about what to block or not based on such analysis rather than on a list of banned sites. A censorship scheme that manages to marry artificial intelligence and basic social-networking analysis would not only be extremely powerful; it would also help to limit the threat that censorship currently poses to economic development, thereby removing one of the major reasons that currently impels governments to avoid censorship.

The Future of Internet Control

The forces that are shaping the future of Internet control come from the realms of politics, society, and business. In the political realm, the U.S. government and its initiatives will be the biggest single force shaping the actions of other governments. Among the key developments to watch will be those concerning the future of the "Internet freedom" agenda and the evolution of the U.S. State Department's approach to the Internet. Hillary Clinton's speech was ambitious and idealistic, but also highly ambivalent. It is unclear how far the State Department is prepared to go in speaking up on behalf of bloggers who are jailed in countries whose rulers serve U.S. interests. Nor is it clear what the broader "Internet freedom" strategy is to be or which projects will receive priority funding. (Some vocal activists from the Middle East have already expressed concerns about the increased U.S.-government funding in this space.)

It remains to be seen whether "Internet freedom" means primarily defending the "freedom of the Internet" (that is, ensuring that governments and corporations avoid increasing censorship and surveillance) or promoting "freedom via the Internet" (that is, using the Internet and new media to facilitate anti-authoritarian movements such as Iran's "Green Wave"). Many governments around the world worry that the latter approach will predominate. Clinton's references to the role that technology played in the protests in Iran (and earlier in Moldova) did nothing to allay those fears.

The tight relationship between the State Department and U.S. technology companies may also prove problematic for both sides, and its future looks uncertain. As European governments and the UN take on "Internet freedom" issues, the State Department may find itself fighting on too many fronts, as those other governments and organizations would probably push to establish new treaties and laws, moves on which Washington is not very keen.

While the State Department promotes a vague notion of "Internet freedom" abroad, a number of domestic law-enforcement and intelligence agencies plus the Commerce Department are pushing for significant changes that amount to "Internet control" initiatives. Taken together, concerns in the areas of cyberwarfare and cybercrime, electronic wiretapping, and Internet piracy and copyright reform may drive the U.S. government toward seeking significant sway over the Internet.

Whatever the democratic merits of such government initiatives, they will have the drawback of creating an enabling environment for authoritarian governments that are keen on passing similar measures, mostly for the purpose of curbing political freedom. In addition, concerns about cybercrime may lead to the proliferation and legitimization of practices such as "deep packet inspection" (when network operators scrutinize the the contents of data packets that pass through their networks), driving down the costs for tools and services associated with it. This, in turn, may abet authoritarian governments (such as Iran's) that are already relying on technology supplied by European companies such as Nokia-Siemens to analyze the traffic passing through national networks.

It is possible that social attitudes toward "publicness" and privacy may become more cautious over time. So far, however, all the indicators are that Internet companies will continue to promote the practice of sharing more and more private data online. Short of U.S. and European policy makers passing new privacy-related legislation—though a few proposals are already in the pipeline—it is unrealistic to expect wider social and cultural shifts away from "publicness." In the business realm, some Internet service providers (ISPs) in Germany and the Netherlands are moving to make DDoS attacks costlier and more difficult to mount by informing any customer whose computer has become infected by a "botnet" (the mass of hijacked computers that makes a DoS attack "distributed"), by requiring corrective measures whenever a botnet infection is detected on a customer's machine, and by urging preventive measures to stop such infections before they start. Absent such interventions, the cost of DDoS attacks will continue to decline as botnets proliferate. Whether all ISPs will accept the potentially expensive task of fighting botnets remains to be seen, however.

Similarly, the software used for analyzing and "mining" data is becoming more powerful as businesses and intelligence agencies demand it. Whether the use of such software could be limited only to democratic states and business contexts remains to be seen; in the worst-case scenario, such tools may end up strengthening the surveillance apparatus of authoritarian states.

Authoritarian governments control the Internet through the combination of technological and sociopolitical means. It is unclear what the most potent combination of those types is; an Internet-control system that wields mainly the sociopolitical means may end up being more draconian than one that relies on technological means only. The great paradox is that the rising profile of "liberation technology" may push Internet-control efforts into nontechnological areas for which there is no easy technical "fix."

Both types of control are made possible by a number of social, political, and technological factors, many of which have their roots in the economies and government policies of democratic states. Any ambitious effort to promote "Internet freedom" should therefore begin by generating a typology of those factors as well as outlining some strategies for dealing with them. The U.S. government's current "Internet freedom" policy has yet to face this challenge, though it needs to do so.

NOTES

1. John Markoff, "Iranians and Others Outwit Net Censors," *New York Times,* 30 April 2009.

2. Virginia Heffernan, "Granting Anonymity," *New York Times Magazine,* 17 December 2010.

3. Charlie Savage, "U.S. Tries to Make It Easier to Wiretap the Internet," *New York Times,* 27 September 2010.

4. Hillary Rodham Clinton, "Remarks on Internet Freedom," 21 January 2010, available at *www.state.gov/secretary/rm/2010/01/135519.htm.*

5. Erica Naone, "Political Net Attacks Increase," *Technology Review,* 13 March 2009.

6. See "Politically Motivated Cyber Attacks," Help Net Security, 6 October 2010, available at *www.net-security.org/secworld.php?id=9957.*

7. "Thai Website to Protect the King," BBC News, 5 February 2009, available at *http://news.bbc.co.uk/2/hi/asia-pacific/7871748.stm.*

8. Peter Burrows, "Internet Censorship, Saudi Style," *Business Week,* 13 November 2008.

9. Soren Billing, "Saudi Campaign to Clean Up YouTube," ITP.net, 13 August 2009, available at *www.itp.net/564689-saudi-campaign-to-clean-up-youtube.*

10. Billing, "Saudi Campaign to Clean Up YouTube."

11. Michael Wines, "In Restive Chinese Area, Cameras Keep Watch," *New York Times,* 2 August 2010.

12. Tom Simonite, "Surveillance Software Knows What a Camera Sees," *Technology Review,* 1 June 2010.

13. Maija Palmer, "Face Recognition Software Gaining a Broader Canvas," *Financial Times,* 22 May 2010.

II

Liberation Technology in China

5

THE BATTLE FOR
THE CHINESE INTERNET

Xiao Qiang

Xiao Qiang is adjunct professor at the Graduate School of Journalism of the University of California–Berkeley, principal investigator at CounterPower Lab, and founder and chief editor of China Digital Times. *This essay originally appeared in the April 2011 issue of the* Journal of Democracy.

Scholars, journalists, and other commentators have extensively explored censorship in the People's Republic of China (PRC), but much remains to be learned. In particular, we need a better grasp of the "cyberpolitics" of expanding online discourse and the capacity of the Internet to advance free speech, political participation, and social change. We also need to know more about the implications of (and limits on) the state's efforts to control what people can see, say, and do online. These issues are crucial to our understanding of China and Chinese society and the role of the Internet under an authoritarian, one-party regime.

It was in 2007—dubbed "Year One of Public Events *(Gonggong Shijian Yuannian)*" by one commentator[1]—that the Internet first helped to propel certain happenings into the official media despite resistance from censors. By doing so, Internet activity effectively set the agenda for public discourse. That year, stories about protests against the Xiamen chemical plant, slave labor at brick kilns, and the abuse of individual property rights spread rapidly online, generating so much public interest and debate that censors and the official media had little option but to report on them as well.

A look at the explosive growth of Internet access and use across China, the tools and methods used by the authorities to control the content and flow of information, and the emerging dynamics between Chinese Internet users, or "netizens," and censors shows that the expansion of the Internet and Web-based media is changing the rules of the game between state and society: Authorities are increasingly taking note and responding to public opinion as it expresses itself online. This trend will surely continue, with online public-opinion formation playing an important role in the future development of Chinese society.

Beginning in March 2007, blogger Lian Yue posted a series of articles warning the people in his hometown, Xiamen in Fujian Province, of the potentially disastrous environmental impact of a proposed paraxylene (PX) chemical factory in the city. He urged his fellow residents to speak out against the plant. Although provincial and city authorities vigorously deleted anti–PX factory messages on servers within their jurisdiction, the offending posts on Lian Yue's blog remained because its server was in another province. Word of the PX plant soon spread throughout the city via e-mail, instant messages (IMs), and text messages on mobile phones. A few months later, in defiance of warnings from local authorities, several thousand people showed up to protest in front of city hall. Participants reported the event live, uploading cellphone photos and texts directly to their blogs. Six months later, following two public hearings on the matter, city authorities decided to relocate the lucrative project. The official Xinhua News Agency praised the turnaround as indicating "a change in the weight given to the views of ordinary Chinese in recent years."[2]

The Xiamen story marks the rise of a remarkable new force in China's contemporary social and political life: popular opinion (communicated online) setting the public agenda together with liberal elements in the traditional media. According to the January 2008 blue book on social development produced by the Chinese Academy of Social Sciences, more than fifty-million Chinese read blogs regularly, making them "an important channel for people to voice their opinions about important events."

The government-run China Internet Network Information Center (CNNIC) found that by the end of 2009, the number of Internet users in China had skyrocketed to 384 million, with 53 million new users going online in the last half of that year alone.[3] CNNIC's 2010 statistics show that users are disproportionately young—more than 60 percent are under 25, and about 70 percent are under 30—and relatively well educated, with more than 40 percent holding postgraduate degrees.[4]

The rise of blogging, instant messaging, social-networking services such as QQ, and search-engine and RSS aggregation tools such as Baidu.com and Zhuaxia.com have given netizens an unprecedented capacity for communication. Internet bulletin-board systems (BBSs)—the primary way in which Chinese netizens access and transmit information online to a large number of people—play a vital role in Chinese online life. By early 2009, China had more than 13 million BBS users, with two-million posts published every day.[5] The Tianya Club *(www.tianya. cn)*, based in Hainan Province, has 33.4 million total registered accounts and between 100,000 and 500,000 users online at any one time. This online community has 200,000 daily online users, hundreds of thousands of new posts, and millions of commentaries a day. Another online forum that is popular among university students, Mop.com, is thought to be even larger, with more than fifty-million visits a day. Users discuss current events by posting comments on the bulletin boards of these major forums as well as in smaller virtual communities. Even when the subject is politically taboo or sensitive, under the cover of anonymity or using coded euphemisms, participants can express particular views—and in far bolder language than would be permitted in the official media.

The "blogosphere" has likewise expanded. Like BBSs, blogs exact only a very low cost of entry—anyone with Internet access can start a blog on a hosting service. According to CNNIC, the number of Chinese with blogs reached 221 million by the end of 2009. Of those, the number of active bloggers had risen to 145 million, a 37.9 percent increase from just six months before.[6]

Although most posts are personal in nature, more and more bloggers are writing about public affairs and becoming local opinion leaders. Blogs usually allow readers to comment, and because they often contain links to other blogs and sites, they act as units in a dynamic community. Together they form an interconnected whole—the blogosphere. While the popular BBSs often become forums where public opinion regarding various topics crystallizes, the redundancy of clusters and links in the blogosphere forms a networked information environment that makes absolute top-down control of content nearly impossible.

In addition to BBSs and blogs, chatrooms and IM services, such as those of QQ and MSN, are also popular channels of communication. On 5 March 2010, Tencent (owner of QQ) announced that the number of simultaneous online QQ users had reached a hundred-million.[7] These IM services play a crucial role in connecting Internet users, facilitating the spread of information, and coordinating actions through social networks. Finally, new photo- and video-sharing sites such as Youku and Tudou are the fastest-growing online applications. The richness of online images, video, and sound has created a powerful media space where millions of users can generate, distribute, and consume content.

Before the Internet, opportunities for unconstrained public self-expression and access to uncensored information were extremely limited. The new online freedoms have developed in spite of stringent government efforts at control and containment. Three Chinese characters may best describe the dynamic between authorities and netizens in Chinese cyberspace: *feng* ("block" or censor), *shai* ("place under the sun" or reveal), and *huo* ("set on fire" or rapidly spread).

Online Censorship

Since the PRC's 1949 founding, information control has been an essential component of the governing strategy of the Chinese Communist Party (CCP). The CCP has a monopoly on political power and has exerted firm control over all mass media, from newspapers and magazines to television channels and radio stations, making them mouthpieces for the Party line. As the reform-minded journalist Lu Yuegang once wrote, the CCP "must depend on two weapons: guns and pens. . . . The logic behind this philosophy is not only to control the pen but to have this control backed by the gun."[8]

Since the introduction of the Internet in China in 1987, the government has employed a multilayered strategy to control and monitor (*feng*) online content and activities. Authorities at various levels use a complex web of regulations, surveillance, imprisonment, propaganda, and the blockade of hundreds of thousands of international websites at the national-gateway level (the "Great Firewall of China").

Several offices govern Internet content—most notably, the CCP's Central Propaganda Department (CPD), which ensures that media and cultural content toe the party line, and the State Council Information Office (SCIO), which oversees all news-providing websites, both official and independent. Municipal, provincial, and county offices of the CPD and SCIO are responsible for overseeing all media published or hosted within their various jurisdictions. CPD officials frequently issue censorship directives to their local counterparts, who have some leeway to implement them as they see fit, and local officials sometimes issue their own censorship directives and fine, threaten, or shut down media outlets that report information which authorities would prefer to keep from the public.[9]

Officials use a number of tactics—keyword filtering, for example—to control online content. The Berkeley China Internet Project obtained a list of more than a thousand words that are automatically banned in China's online forums, including *dictatorship, truth,* and *riot police.* Censors customarily do not make clear exactly what content they intend to ban. The government's primary strategy for shaping content is to hold Internet service providers (ISPs) and access providers responsible for the behavior of their customers; thus business operators have little choice but to censor the content on their sites proactively. For example, regulations posted by the Guangdong Provincial Communications Administration state:

> The system operator will be responsible for the contents of his/her area, using technical means as well as human evaluation to filter, select, and monitor. If there should be any content in a BBS area that is against the regulations, the related supervisory department will hold the BBS as well as the individual operator responsible.[10]

Business owners must use a combination of their own judgment and direct instructions from propaganda officials to determine what content to ban. In an anonymous interview with this author, a senior manager at one of China's largest Internet portals acknowledged receiving instructions from either SCIO or other provincial-level propaganda officials at least three times a day.

Additionally, both the government and numerous websites employ people to read and censor content manually. Tens of thousands of websites hosted overseas are also blocked at the level of the nine national gateways that connect the Chinese Internet to the global network.[11] Websites hosted inside China can be warned or shut down if they violate rules of acceptable content, and individual Internet users who spread information that authorities deem harmful have been threatened, intimidated, or thrown in jail, most often on charges related to national security, such as "subversion." Speaking to the CCP Politburo in January 2007, President Hu Jintao called for improved technologies, content controls, and network security for monitoring the Internet, saying, "Whether we can cope with the Internet is a matter that affects the development of socialist culture, the security of information, and the stability of the state."[12]

Of course the government was already proving creative in its policing of the Internet. For example, since 2007 two cartoon characters, Jingjing and Chacha (from *jingcha,* the Chinese word for police), have popped up on Internet users' screens to provide links to the Internet Police section of the Public Security website, where readers can report illegal online information. A Shenzhen police officer explained: "This time we publish the image of Internet Police in the form of a cartoon, to let all Internet users know that the Internet is not a place beyond the law . . . The main function of Jingjing and Chacha is to intimidate, not to answer questions."[13]

Throughout 2008, Internet control was increasingly tightened in order to present a harmonious image to the world during the Beijing Olympics. Beginning in early 2009, the government further ratcheted up its efforts at control. The initiatives included a campaign against "vulgarity" (which encompasses not just pornography, but also dirty words, slang, and socially and politically unacceptable images) that aimed at search engines, Web-hosting services, and online communities.[14] According to official Chinese media reports, thousands of websites were closed as a result.

Likewise, as the twentieth anniversary of the Tiananmen Square massacre neared that year, the government temporarily shut down countless websites—including Facebook, Twitter, and Wikipedia—ostensibly for "technical maintenance." Then, right around the anniversary on June 4, the Ministry of Information Technology announced plans to require the preinstallation of a filtering software called Green Dam Youth Escort on all computers made or sold in China. After public outcry, however, these plans were scrapped. On July 5, in the aftermath of interethnic riots in Urumqi, Xinjiang Province—home to most of China's Uyghur population—the government again blocked Twitter and other microblogging sites.

Even with the censors' constant presence, the ephemeral, anonymous, and networked nature of Internet communication limits their impact. Moreover, a number of factors make the censors' work particularly daunting. First, the Internet is a many-to-many communication platform that has very low barriers to entry (and risks of use) for anyone who has an Internet connection. Second, with the complicated network topology—the blogosphere and the whole Internet with its redundant connections, millions of overlapping clusters, self-organized communities, and new nodes growing in an explosive fashion—traditional methods of content control such as putting pressure on a publisher to self-censor become nearly impossible.

The Chinese government's Internet-control system mainly aims to censor content that openly defies or attacks CCP rule or contradicts the official line on such taboo topics as the Tiananmen Square massacre or Tibet. Most important, however, is preventing the widespread distribution of information that could lead to collective action such as mass demonstrations or signature campaigns.

The Digital Resistance

The results of government censorship efforts are mixed at best. In early 2009, a creature named the "Grass Mud Horse" appeared in an online video that became an immediate Internet sensation.[15] Within weeks, the Grass Mud Horse—or *cao ni ma,* the homophone of a profane Chinese expression—became the de facto mascot of Chinese netizens fighting for free expression. It inspired poetry, videos, and clothing lines. As one blogger explained, the Grass Mud Horse represented information and ideas that could not be expressed in mainstream discourse.

The Grass Mud Horse was particularly suited to the contested space of the Chinese Internet. The government's pervasive and intrusive censorship has stirred resentment among Chinese netizens, sparking new forms of social resistance and demands for greater freedom of information and expression, often conveyed via coded language and metaphors adopted to avoid the most obvious forms of censorship. As a result, the Internet has become a quasi-public space where the CCP's dominance is exposed, ridiculed, and criticized, often by means of satire, jokes, songs, poems, and code words.

Such coded communication, once whispered in private, is not new to China. Now, however, it is publicly communicated rather than murmured behind the backs of the authorities. For example, since censorship is carried out under the official slogan of "constructing a harmonious society," netizens have begun to refer to the censoring of Internet content as "being harmonized." Furthermore, the word "to harmonize" in Chinese *(hexie)* is a homonym of the word for "river crab." In folk language, *crab* also refers to a bully who exerts power through violence.

Thus the image of a crab has become a new satirical, politically charged icon for netizens who are fed up with government censorship and who now call themselves the River Crab Society. Photos of a malicious crab travel through the blogosphere as a silent protest under the virtual noses of the cyberpolice. Even on the most vigorously self-censored Chinese search engine, Baidu.com, a search of the phrase "River Crab Society" will yield more than 5.8 million results.

In China, the nebulous nature of the Internet allows information not easily accessible elsewhere to be revealed *(shai)*. Anyone who goes on-line will be exposed to multifarious sources of information and have unprecedented opportunities to exchange ideas and opinions on social, political, and personal issues. Simultaneously, the interaction between information and communications technology and the traditional media creates a dynamic that is challenging the boundaries of the existing censorship system and thereby the official media as well.

Netizens have launched endless so-called *shai* activities on bulletin boards, blogs, and video- and photo-sharing services: For "*shai* salaries," people post their own or others' salaries for comparison; for "*shai* vacations," users share vacation photos and experiences; and for "*shai* corruption," "*shai* bosses," and "*shai* riches," netizens publish information and opinions online about the elite that would otherwise go unsaid.

The *feng* and *shai* processes are constantly at odds with each other. Even when information is censored at a high level, it often makes its way through online cracks to spread among netizens. In addition, foreign websites and news media that provide Chinese-language services—including the BBC, Radio Free Asia (RFA), and newspapers based in Hong Kong and Taiwan—frequently publish information that is censored in China, which is then often redistributed inside the country by a small but active group of tech-savvy "information brokers" who know how to circumvent the Great Firewall and circulate the news via BBSs, mass e-mailings, and other online channels. Thus banned publications such as dissident newsletters and Voice of America updates can reach Chinese readers despite the government's use of advanced filtering technology.

The last character, *huo* ("fire" or information cascade), describes the phenomenon of news reports, comments, photos, or videos that spread through cyberspace like wildfire. The original item may appear on a bulletin board or blog post, or even in a local paper, and can generate thousands of subsequent comments and posts. Like water gushing through a hole in a dam, if the speed and volume of the dissemination is great enough, any attempt to stanch the flow will be in vain. Driven by dense clusters of interested netizens, the spread of controversial information can outpace the control of censors and quickly become public knowledge—a state of affairs that has huge political implications. When a *huo* phenomenon occurs, the Internet plays the role of a massive distribution platform that denies the government its agenda-setting power.

The *huo* process is especially potent when a local issue resonates with a broader audience and spreads beyond the limited jurisdiction of local officials, sometimes even making it into the national media. When corruption or environmental damage, for example, are exposed, local authorities implicated in the scandal often crack down on news websites hosted within their respective jurisdictions. But when such news finds its way to a website based outside the relevant local jurisdiction, the officials of that jurisdiction will have no means of directly suppressing it and no guarantee that their counterparts in other locales will choose to do so. Central authorities may likewise choose not to impose a blackout of online news about a problematic local issue or event. This gap in control between local authorities as well as between local and central authorities opens a space for netizens to transmit information.

For example, when drunken 22-year-old Li Qiming ran down two roller-blading college students on the campus of Hebei University— killing one and injuring the other—and was arrested after leaving the scene, he shouted: "Go ahead, sue me if you dare. My dad is Li Gang!" Li Gang was the deputy director of the local public-security bureau. Four days later, Mop.com ran an online contest asking entrants to incorporate the sentence "My father is Li Gang" into classical Chinese poems. The contest garnered more than six-thousand entries. A few days after that, the CPD issued a directive to prevent any further "hype regarding the disturbance over traffic at Hebei University." But the phrase "my father is Li Gang" has since become a popular Internet meme in China.[16]

The *huo* phenomenon also plays a critical role in the interplay between Internet expression and changes in the traditional media. Many Chinese journalists are leading double lives—reporters for the state-controlled media by day, bloggers by night. When covering touchy subjects—such as natural disasters, major industrial accidents, or official-corruption cases—print reporters must follow the lead of official sources before conducting interviews and publishing their findings. But journalists can now evade such guidelines by collecting and distributing information online, making it harder for censors to hush up sensitive stories. In fact, when such information becomes *huo* online, the traditional media have a legitimate reason to cover it. Some even follow breaking developments as these are reported in the online realm. *Southern Weekend,* for example, has an editorial section called Net Eye that picks up interesting online stories and publishes them in print.

An Avenue for Feedback and Accountability

In recent years, the processes of *feng, shai,* and *huo* have been at work, helping the Internet to drive public opinion. In early 2007, a netizen from Chongqing posted a photograph of a house, dubbed "China's Most Incredible Nail House," being threatened by a new development.[17] The image touched on problems of urban construction, property rights, and forced evictions, and the official media soon jumped on the story, which happened to break just as the National People's Congress was passing a new property-rights law that purported to protect individual homeowners.

Even as the official media began to carry the story, Sina.com (China's largest Internet portal) offered to pay for images and videos of the nail house, and Mop.com ran a real-time monitoring page. A local court ruled against the homeowners, but public opinion, swayed by poignant online images, heavily favored the nail-house owners. After they disobeyed the court order and refused to move, the central government issued orders to limit reporting, but the story lived on through photographs posted online by netizens. Ultimately, the developer bowed to public pressure, settling the case and compensating the couple for their property, which was eventually destroyed.

Sometimes the government has official reasons to acknowledge certain elements of a story while censoring others, as with news of the widespread use of slave labor at brick kilns in Shanxi Province in mid-2007. The story, which involved the kidnapping of children, slave labor, and the collusion of local police, Party officials, and kiln owners, spread through the Chinese blogosphere and ignited national outrage. Reports in the official media followed, and soon top Party officials—including President Hu Jintao and Premier Wen Jiabao—publicly expressed concern over the issue. Details continued to emerge, and the story only got uglier, eventually spurring the Internet Bureau of the SCIO (also called the CCP External Communication Office) to instruct its subordinate offices and the main Internet news portals to stick to "positive propaganda," to emphasize the government's responsiveness in the brick-kiln matter, to ramp up the monitoring of websites, and to quickly delete information that could be harmful to the government.[18]

Although these cases show that official Internet censorship is not always automatically and fully employed nor always successful, in general the government is able to exert near-total control over information distributed online, particularly when officials make this control a priority. For example, when dissident writer Liu Xiaobo was awarded the 2010 Nobel Peace Prize, the CPD ordered all websites not to create or post stories about the prize and to delete any that already existed. The SCIO also issued a directive forbidding all interactive online forums, including blogs and microblogs, from transmitting prohibited words relating to the prize.[19] As a result, hardly any mention of last year's Nobel Peace Prize can be found on the Chinese Internet, let alone Liu Xiaobo's name or writings.

Nonetheless, there is a changing dynamic afoot: Some big stories are breaking online first only to be carried later by traditional media, thereby making bloggers and netizens information agenda setters. Moreover, despite government censorship efforts, the sheer speed and number of messages and Internet posts are making it ever harder, and in some cases impossible, for censors to stay ahead. The time lapse between the information cascade and top-down censorship instructions is critical, as is the gap in control between central and local authorities, which has allowed local events to become national news reported by the centrally controlled media. Once sensitive stories appear in the official media, the Internet amplifies and keeps them alive, preventing the government from ignoring or suppressing inconvenient news.

Since traditional media outlets still remain under CCP control, even the more progressive and outspoken publications such as *Southern Metropolis Daily* or *Southern Weekend* have only a very limited ability to push the envelope on political reporting. When mass protests, health epidemics, or official-corruption cases occur, the Internet is now the first place people go to find the latest news and to share experiences and opinions. For the first time, citizens are able to participate in a public dialogue about issues of crucial importance to their lives.

In 2007, Xinhuanet surveyed the most popular topics (not including those that had been deleted by monitors) in the three most influential online communities—Strong Country Forum *(bbs.people.com.cn)*, Tianya Club, and Kaidi *(club2.cat898.com)*.[20] In addition to the big stories that year, the study also found that other "sensitive social events" were popular, including those relating to governance, police violence, environmental protection, public health, judicial reform, and natural disasters. It also suggested that netizens' consciousness of rights is rising, as expressions such as "right to know," "right to express," and "right to monitor [the government]" are often used in connection with those large online public events. Furthermore, participants cited as concerns the credibility and responsiveness of various levels of government as well as issues of public morality and the crisis of values in society.

As Beijing-based Internet expert Hu Yong has written: "Since China never had mechanisms to accurately detect and reflect public opinion, blogs and BBS have become an effective route to form and communicate such public opinions of the society."[21] One of the direct impacts of this new information landscape is that negative reports and criticism of local officials—especially relating to corruption, social justice, or people's daily experiences—are now being exposed and nationally disseminated online and resonating across society. Sometimes such a process is tolerated by central authorities to keep lower officials in check and to allow the public to let off steam before it erupts uncontrollably, perhaps resulting in public protests. Such Internet-generated public opinion is sometimes the sole channel for providing feedback to officials.

Online oversight has an especially large impact on local officials in charge of administrative, legal, law-enforcement, and propaganda agencies. Once local officials lose control and information spreads beyond their jurisdiction, the vehemence of an aroused public may force them to change policies. In the words of a local propaganda official, for the government, "it was so much better when there was no Internet."[22]

Internet-driven public events have also helped to highlight issues that originate locally—or even abroad—but have wider implications for Chinese society. For individuals advocating political reform and social change, the Internet and the more reform-minded parts of the traditional media offer outlets to discuss topics that had before been taboo. For example, as prodemocracy protests erupted in Egypt in January 2011, Chinese authorities ordered that only Xinhua's account of events be disseminated. Yet even as Sina Weibo, China's version of Twitter, blocked the word "Egypt" from its search engine,[23] hundreds of thousands of posts on Sina Weibo remained available to savvy users, and netizens continued to spread the news from Egypt and discuss its implications for China's political reality. Eventually, in the wake of online calls for prodemocracy demonstrations inspired by events in Tunisia and Egypt, the word "Jasmine" was also barred from Sina Weibo.[24]

Citizen Mobilization

Often, the next step after public dialogue is collective mobilization and organization around issues of common concern. This is an area where Internet-based public opinion has the potential to make a powerful impact on Chinese society and politics. While authorities stifle civil society and independent social organizations, various grassroots groups that depart from the official line with regard to such social issues as the environment, women's rights, and homosexuality rely on the Internet to organize and distribute information. The expanded space for discussion of public affairs has allowed civil society to push the boundaries of associative and communicative freedoms.

The Xiamen anti-PX protests are now considered a milestone. One protester told a foreign reporter covering the story that at last, the people "can be heard." The city government, in return, listened to public opinion and adjusted its decision accordingly. This was a first in China and a very encouraging sign. The state-run Xinhua News Agency concluded, "The suspended controversial Xiamen city PX plant probably will not become a landmark wherever it finally stands, but it may have helped lay a cornerstone that boosts ordinary Chinese people's participation in policy making."[25] Of course the ruling CCP has not shown any sign of giving up its monopoly on political power and is still highly sensitive to the growing political impact of the Internet.

Online mobilization and protests have also made an impact beyond China's borders, becoming a significant factor influencing Chinese diplomacy and the country's image abroad. In November 2009, to commemorate the twentieth anniversary of the fall of the Berlin Wall, a German nonprofit created a virtual "Berlin Twitter Wall" where indi-

viduals could post their thoughts on the occasion through use of the
Twitter hashtag "#FOTW." The site's introduction further invited par-
ticipants to "let us know which walls still have to come down to make
our world a better place!" In response, Chinese comments blasting the
Great Firewall and Internet censorship dominated the virtual Berlin Wall
for weeks. Alluding to Ronald Reagan's famous speech before Berlin's
Brandenburg Gate, Chinese bloggers also waged a "Tear Down This
Firewall!" campaign prior to U.S. president Barack Obama's visit to
China in November 2009. Largely due to such efforts, President Obama
addressed the issue of online freedom of speech at a town-hall meeting
with students in Shanghai.

Not all online mobilization is as spontaneous and anonymous as the
campaign against the Great Firewall was. Influential bloggers may also
mobilize their fellow netizens by acting as spokespersons for certain
issue positions, or by giving personal authentication to messages that
resonate with the people, or by articulating what others could not say
in the face of political censorship. Bestselling author, race-car driver,
and blogger Han Han is one such figure. Han is an outspoken critic of
government censorship, and his blog posts are often deleted by censors.
Nevertheless, his main blog received more than three-hundred million
hits between 2006 and 2009. In April 2010, *Time* magazine listed Han
Han as a candidate for the hundred "most globally influential people."
Han Han subsequently wrote a blog post asking the Chinese government "to treat art, literature, and the news media better, not to impose
too many restrictions and censorship, and not to use the power of the
government or the name of the state to block or slander any artist or
journalist."[26] This post generated some 25,000 comments from his readers and was viewed by more than 1.2 million people. The article has
also been widely reposted online; in May 2010, a Google search found
more than 45,000 links reposting all or part of the essay. Despite official efforts to use the Great Firewall to block Chinese netizens from
voting for Han Han on *Time*'s website, he came in second in the final
tally, showing the mobilizational power of his writing.

Government officials have begun to recognize that the Internet has
set an irreversible trend toward a society that is more transparent, a citi-
zenry that is more eager to participate in public life, and a public whose
opinion carries more weight. Some officials advocate the need for political reform to adapt to these forces. In a long 2007 article published
in the official press, Xin Di, director of the Research Department of the
Central Party School, listed five concrete examples to show the "incre-
mental progress" taking place in China's political system.[27] Four of his
five examples were not top-down efforts at political reform, but rather
were related to official government reactions to Internet-driven public
events. Although genuine political reform did not appear on the leadership's agenda at the CCP National Congress that year, some lower-level

officials did recognize the important role of the Internet as a catalyst for political change in Chinese society.

These more forward-looking officials believe that the government should selectively tolerate or even welcome Internet expression as a barometer for public opinion. Permitting such expression allows the government to collect information about society, to be more responsive to citizens' concerns, and to provide a safety valve for the release of public anger. The Internet can also help to hold local officials more accountable—to the central authorities as well as to the public. In addition, the Internet plays a role in promoting political change when the interests and agendas of different government agencies or administrative levels do not align. In such a case, public opinion may help to bolster one side over the other.

A 2009 Chinese Academy of Social Sciences study of the Internet's impact on public opinion identifies netizens as a "new opinion class" that can swiftly influence society and describes the dual methods used by the government to cope with the growing challenge of online activism— clamping down on the Internet while also responding quickly to public opinion.[28] Indeed, this rising online public participation is an indicator that the rules of the political game in China may have started to change.

The CCP's censorship of both the traditional media and the Internet is certain to continue. Yet the increasing influence of online public opinion shows that the CCP and the government can no longer maintain absolute control over the spread of information. The Internet is already one of the most influential media spaces in China—no less so than traditional forms of print or broadcast media. Underlying ever-stronger measures aimed at bolstering state control is a rising level of public information and awareness in Chinese society. Furthermore, through online social networks and virtual communities, the Chinese Internet has become a substantial communications platform for aggregating information and coordinating collective action.

The conflicting forces of *feng, shai,* and *huo* will remain in tension with one another. The result is an emerging pattern of public opinion and citizen participation that represents a shift of power in Chinese society. The Internet allows citizens to comment on certain (albeit limited) topics, and to move them out of purely local arenas to the point where they can become national concerns. Moreover, these "public events" now play a role in promoting human rights, freedom of expression, the rule of law, and governmental accountability. An entire generation of online public agenda setters has emerged to become influential opinion leaders. Surely they will have an important role to play in China's future.

Furthermore, certain somewhat progressive media outlets such as *Southern Metropolis Daily* and *Southern Weekend* are also actively expressing more liberal political ideas and pushing the envelope whenever they can. Before the Internet, such reform-minded discourse could gain little ground against CCP propaganda. Now, as these liberal elements within the established media converge with independent, grassroots voices online, they are creating a substantial force that is slowly wearing away at the CCP's ideological and social control.

China is becoming an increasingly transparent and mobile society with more pluralistic values. The Internet has become a training ground for citizen participation in public affairs: It creates a better informed and more engaged public that is demanding more from its government. The CCP regime is learning to adapt to these new circumstances and becoming more responsive. Already we are starting to see compromise, negotiation, and rule-changing behavior in the regime's response to this challenge, indicating the possibility of better governance with greater citizen participation. From this perspective, the Internet is not just a contested space, but a catalyst for social and political transformation.

NOTES

1. "Xiao Shu: Hoping Xiamen PX Event Became the Milestone," *Nanfang Zhoumo* [Southern Weekend] (Guangzhou), 20 December 2007.

2. See Xinhuanet at *http://news.xinhuanet.com/comments/2007-12/23/content_7297065. htm.*

3. China Internet Network Information Center (CNNIC), "25th Statistical Report on the Development of China's Internet," January 2010; available at *www.cnnic.net.cn/upload-files/pdf/2010/1/15/101600.pdf.*

4. CNNIC, "25th Statistical Report," and CNNIC, "20th Statistical Report on the Development of China's Internet," June 2007. Since July 2007, CNNIC has counted as "Internet users" anyone over six years old who visited the Internet from any terminal (including a mobile phone), at least once in six months. Before July 2007, an Internet user was defined as anyone who spent at least an hour a week online. As a result of this change, more low-income users, such as migrant workers and rural residents, who use mobile phones instead of personal computers as their main communications interface, are now being included as netizens.

5. "New Media Clash," Xinhua News Agency, 16 February 2009, available at *http://news.xinhuanet.com/zgjx/2009-02/16/content_10825818.htm.*

6. CNNIC, "25th Statistical Report."

7. "Tencent Announces QQ Users Reached 100 Million," 5 March 2010, see *http://tech.qq.com/a/20100305/000528.htm.*

8. "A Bold New Voice—Lu Yuegang's Extraordinary Open Letter to Authorities," *China Digital Times,* 20 July 2004.

9. U.S.-China Economic and Security Review Commission, Hearing on Access to Information in the People's Republic of China, testimony by Xiao Qiang, 31 July 2007.

10. U.S.-China Economic and Security Review Commission, Hearing on China's State Control Mechanisms and Methods, testimony by Xiao Qiang, 14 April 2005.

11. Jonathan Zittrain and Benjamin Edelman, "Empirical Analysis of Internet Filtering

in China," working paper, Berkman Center for Internet and Society, Harvard Law School, 2003; available at *http://cyber.law.harvard.edu/filtering/china.*

12. "Hu Jintao Asks Chinese Officials to Better Cope with Internet," *People's Daily,* Beijing, 24 January 2007.

13. "Starting from September 1, New Virtual Cops Will 'Cruise' All Thirteen Internet Portals in Beijing," *Beijing News,* 22 August 2007.

14. The other agencies are the State Administration for Industry and Commerce (SAIC), State Administration of Radio, Film and Television (SARFT), and General Administration of Press and Publication (GAPP).

15. See *http://chinadigitaltimes.net/2009/02/music-video-the-song-of-the-grass-dirt-horse.*

16. Michael Wines, "China's Censors Misfire in Abuse-of-Power Case," *New York Times,* 17 November 2010, available at *www.nytimes.com/2010/11/18/world/asia/18li.html.*

17. See *www.flickr.com/photos/scorpico7/2765449045* for a photograph of the nail house.

18. "A Notice from the Central Government to Censor News Related to Shanxi Brick Kilns Event," *China Digital Times,* 15 June 2007, available at *http://chinadigitaltimes.net/2007/06/a-notice-from-the-central-government-to-censor-news-related-to-shanxi-brick-kilns-event.*

19. "New Directives from the Ministry of Truth (RE: Liu Xiaobo Wins Nobel Peace Prize)," *China Digital Times,* 8 October 2010, available at *http://chinadigitaltimes.net/2010/10/new-directives-from-the-ministry-of-truth-october-8-2010-re-liu-xiaobo-wins-nobel-peace-prize.*

20. "Study Report of Online Public Opinions in 2007," Xinhuanet, 5 February 2008.

21. Hu Yong, "Blogs in China," China Media Project Case Study (on file at the Journalism and Media Studies Centre, University of Hong Kong), 4 August 2005.

22. "Director of Propaganda Department of Shuide, Shanxi Province: Those Years Without Internet Were So Much Better!" Southern News.net, 30 January 2008, available at *http://news.qq.com/a/20080130/000639.htm.*

23. Pascal-Emmanuel Gobry, "China Blocks 'Egypt' on Sina Weibo, Its Twitter Clone," Business Insider SAI, 29 January 2011, available at *www.businessinsider.com/china-blocks-egypt-on-sina-weibo-its-twitter-clone-2011-1.*

24. Kathrin Hille, "China Authorities Block Democracy Campaigns," *Financial Times,* 25 February 2011.

25. "Common Chinese Have More Say in Policy-Making," Xinhuanet, 3 January 2008.

26. See "Han Han . . . Comes in at Number Two in Time 100 Poll: 'Let the Sunshine In,'" *China Digital Times,* 29 April 2010, available at *http://chinadigitaltimes.net/2010/04/han-han-let-the-sunshine-in.*

27. "Political Civilization in Detail," Xinhua News Agency, 4 February 2008.

28. Zhu Huaxin, Shan Xuegang, and Hu Jiangchun, "2009 China Internet Public Opinion Analysis Report," in Chinese Academy of Social Sciences, "2010 Society Blue Paper," 22 December 2009.

6

CHINA'S "NETWORKED AUTHORITARIANISM"

Rebecca MacKinnon

Rebecca MacKinnon *is a Bernard L. Schwartz Senior Fellow at the New America Foundation. She is cofounder of Global Voices Online (www.globalvoicesonline.org), a global citizen-media network. This essay, which draws on testimony that she gave before the U.S. Congressional-Executive Commission on China (www.cecc.gov) on 24 March 2010, originally appeared in the April 2011 issue of the* Journal of Democracy.

To mark the twentieth anniversary of the fall of the Berlin Wall, a German arts organization launched a website called the "Berlin Twitter Wall." Anyone anywhere on the Internet could use Twitter to post a comment into one of the speech bubbles. Within a few days of its launch, the website was overrun by messages in Chinese. Instead of talking about the end of the Cold War and the fall of communism in Europe, Chinese Twitter users accessed the site to protest their own government's Internet censorship. One wrote: "My apologies to German people a million times [for taking over this site]. But I think if Germans learn about our situation, they would feel sorry for us a million times." Twitter is blocked in China. Still, a growing community is so determined to gain access to the widely used social-networking service and hold uncensored conversations with people around the world that these Chinese Internet users have acquired the technical skills to circumvent this censorship system—widely known as the "Great Firewall of China," a filtering system that blocks websites on domestic Internet connections.

In late January 2010, U.S. secretary of state Hillary Clinton—who two months earlier had stood at Berlin's Brandenburg Gate with other world leaders to celebrate the twentieth anniversary of the fall of the Wall—gave a 45-minute speech on "Internet Freedom." She spelled out how one single, free, and open global Internet is an essential prerequisite for freedom and democracy in the twenty-first century. "A new information curtain is descending across much of the world," she warned.

"And beyond this partition, viral videos and blog posts are becoming the *samizdat* of our day."[1]

But can we assume that Chinese authoritarianism will crumble just as the Iron Curtain crumbled two decades ago? It is unwise to make the assumption that the Internet will lead to rapid democratization in China or in other repressive regimes. There are difficult issues of government policy and corporate responsibility that must be resolved in order to ensure that the Internet and mobile technologies can fulfill their potential to support liberation and empowerment.

When an authoritarian regime embraces and adjusts to the inevitable changes brought by digital communications, the result is what I call "networked authoritarianism." In the networked authoritarian state, the single ruling party remains in control while a wide range of conversations about the country's problems nonetheless occurs on websites and social-networking services. The government follows this online chatter, and sometimes people are able to use the Internet to call attention to social problems or injustices and even manage to have an impact on government policies. As a result, the average person with Internet or mobile access has a much greater sense of freedom—and may feel that he has the ability to speak and be heard—in ways that were not possible under classic authoritarianism. At the same time, in the networked authoritarian state, there is no guarantee of individual rights and freedoms. Those whom the rulers see as threats are jailed; truly competitive, free, and fair elections are not held; and the courts and the legal system are tools of the ruling party.

As residents of a networked authoritarian society, China's more than four-hundred million Internet users are managing to have more fun, feel more free, and be less fearful of their government than was the case even a mere decade ago. At the same time, however, the government has continued to monitor its people and to censor and manipulate online conversations to such a degree that no one has been able to organize a viable opposition movement. According to the Dui Hua Foundation, a human-rights advocacy organization, arrests and indictments on charges of "endangering state security"—the most common charge used in cases of political, religious, or ethnic dissent—more than doubled in 2008 for the second time in three years.[2] Average Chinese citizens, however, rarely hear of such trends—an "information gap" which makes it much less likely that a critical mass of them will see the need for rapid political change. The system does not control all of the people all of the time, but it is effective enough that even most of China's best and brightest are not aware of the extent to which their understanding of their own country—let alone the broader world—is being blinkered and manipulated. All university students in China's capital now have high-speed Internet access. But when a documentary crew from U.S. public television recently went onto Beijing university campuses and showed students the iconic 1989 photograph of a man standing in front of a tank in Tiananmen Square, most did not recognize the picture at all.

The Chinese experience teaches us a globally applicable lesson: Independent activists and prodemocracy movements may have won some early skirmishes against censorship, but one cannot assume that their adversaries will remain weak and unskilled in the navigation and manipulation of digital communications networks. In fact, governments and others whose power is threatened by digital insurgencies are learning quickly and pouring unprecedented resources into building their capacity to influence and shape digital communications networks in direct and indirect ways. As Larry Diamond put it: "It is not technology, but people, organizations, and governments that will determine who prevails."[3]

In the public discourse about the Internet and repressive regimes, Western policy makers and activists frequently use Cold War–era metaphors in ways that are similar to Clinton's likening of blogs to Soviet-era *samizdat*. Such metaphors are strongest in the policy discourse about the Great Firewall of China. The Hong Kong–based communications scholar Lokman Tsui has criticized this "Iron Curtain 2.0" lens through which many in the West seek to understand the Chinese government's relationship with the Internet. "Strategies to break down the Great Firewall," he writes, "are based on the belief that the Internet is a Trojan Horse (another metaphor!) that eventually will disempower the Chinese state from within and topple the authoritarian government, as the barbarians in previous times have done for China, and as international broadcasting has done with regard to ending communism in the Cold War." Tsui argues that this framework for understanding the impact of the Internet on Chinese politics is not consistent with the growing body of empirical research and is therefore likely to result in failed policy and activism strategies.[4]

Guobin Yang, who began researching Chinese online discourse even before the Internet first became commercially available there in 1995, has concluded that in spite of China's increasingly sophisticated system of censorship and surveillance, the Chinese Internet is nonetheless a highly "contentious" place where debate is fierce, passionate, and also playful. After analyzing numerous cases in which Chinese Internet users succeeded in bringing injustices to national attention or managed to cause genuine changes in local-government policies or official behavior, Yang argues that the Internet has brought about a "social revolution, because the ordinary people assume an unprecedented role as agents of change and because new social formations are among its most profound outcomes."[5] Note that the revolution he describes is being waged mainly by Chinese people posting and accessing information on websites and services operated by Chinese companies—in other words, acting *inside* the Great Firewall.

In examining the use of information and communications technologies (ICTs) by China's "have-less" working classes, Jack Linchuan Qiu documents how Internet and mobile-phone use has spread down to the "lower strata" of Chinese society. This development has given birth to a new "working-class network society" that provides China's less fortunate people with tools for mobility, empowerment, and self-betterment. Yet he also describes how "working-class ICTs" provide new levers for government and corporations to organize and control a new class of "programmable labor." While Chinese workers have been able to use Internet and mobile technologies to organize strikes and share information about factory conditions in different parts of the country, Qiu concludes that "working-class ICTs by themselves do not constitute a sufficient condition for cultural and political empowerment."[6]

Can Online Activism Help Authoritarians?

In his book *Technological Empowerment: The Internet, State, and Society in China,* Yongnian Zheng points out that the success or failure of online activism in China depends on its scope and focus, and that some online activism—particularly that which is at the local level or targets specific policy issues over which there are divisions or turf wars between different parts of the government—can actually serve to bolster regime legitimacy. The least successful online movements tend to be those that advocate various forms of political "exit," including calls for an end to one-party rule by the Chinese Communist Party (CCP) and greater political autonomy or independence for particular ethnic or religious groups. "When the regime is threatened by challengers," Zheng writes, "the soft-liners and hard-liners are likely to stand on the same side and fight the challengers." On the other hand, successful online movements in China are usually characterized by what Zheng (following Albert O. Hirschman) calls the "voice" option, or what other political scientists call the "cooperation option." Such online insurgencies actually provide ammunition to reformist leaders or liberal local bureaucrats in their power struggles against hard-line conservative colleagues. Voice activism helps reduce political risks to reformist officials, who can point to online sentiment and argue that without action or policy change there will be more unrest and public unhappiness.[7]

Thus, rising levels of online activism in China cannot automatically be interpreted as a sign of impending democratization. One must examine what kind of online activism is succeeding and what kind is failing. If voice activism is for the most part succeeding while exit activism is systematically being stifled and crushed—thanks to high levels of systematic censorship and surveillance, in addition to the lack of an independent or impartial judiciary—one can conclude that the CCP has adapted to the Internet much more successfully than most Western observers realize. The Iron Curtain 2.0 mentality criticized by Tsui may indeed have blinded many Western policy makers, human-rights activists, and journalists to what is really happening in China. In 2005, *New York Times* columnist Nicholas Kristof wrote breathlessly: "it's the Chinese leadership itself that is digging the Communist Party's grave, by giving the Chinese people broadband."[8] Zheng's analysis, however, supports the opposite conclusion: The Internet may actually prolong the CCP's rule, bolstering its domestic power and legitimacy while the regime enacts no meaningful political or legal reforms.

Public-policy discourse and deliberation are not exclusive features of democracies. Political scientists have identified varying amounts of public discourse and deliberation in a range of authoritarian states. In 2008, Baogang He and Mark Warren coined the term "authoritarian deliberation" to explain how China's authoritarian regime uses "deliberative venues" to bolster regime legitimacy. While it is possible that the deliberation now taking place within Chinese authoritarianism might bring about eventual democratization, Baogang He and Warren believe that this is only one of two possibilities. The other is that the deliberative practices embraced by the state could stabilize and extend the CCP's authoritarian rule.[9]

Min Jiang applies the concept of authoritarian deliberation specifically to Chinese cyberspace, identifying four main deliberative spaces: 1) "central propaganda spaces," meaning websites and forums built and operated directly by the government; 2) "government-regulated commercial spaces," meaning websites and other digital platforms that are owned and operated by private companies but subject to government regulation, including elaborate requirements for content censorship and user surveillance; 3) "emergent civic spaces," meaning websites run by nongovernmental organizations and noncommercial individuals, which are censored less systematically than commercial spaces but are nonetheless subject to registration requirements as well as intimidation, shutdown, or arrest when authors cross the line or administrators fail to control community conversations; and 4) "international deliberative spaces," meaning websites and services that are hosted beyond Chinese-government jurisdiction—some of which are blocked and require circumvention tools to access—where content and conversations not permitted on domestic websites can be found, and where more internationally minded Chinese Internet users seek to conduct conversations with a broader global public.

It is important to note that the Great Firewall is meant to control only the fourth category of deliberative space, the one that is located outside China. Yet it is the first two categories, as Jiang points out, that have the greatest impact on Chinese public opinion. The state uses much more direct and proactive means to control the first three deliberative spaces, all of which operate within the jurisdiction of the Chinese government. Undesirable or "sensitive" content is either deleted from the Internet altogether or blocked from being published.[10]

The Web as Waterworks

Chinese scholar Li Yonggang has suggested that, instead of using a "firewall" metaphor, it is more helpful to think of Chinese Internet controls—which include not only censorship but surveillance and manipulation of information—as something like a hydroelectric water-management system. Managers have both routine and crisis-management goals: managing daily flows and distribution on the one hand and managing droughts and floods on the other. It is a huge, complex system with many moving parts, and running it requires flexibility. It is impossible for the central government to have total control over every detail of water level or pressure at any given time. The system's managers learn and innovate as they go along.[11]

Recent Chinese-government statements show that, like water, the Internet is viewed as simultaneously vital and dangerous. According to the 2010 government white paper "The Internet in China," rapid, nationwide expansion of Internet and mobile-device penetration is a strategic priority. The Internet is seen as indispensible for education, poverty alleviation, and the efficient conveyance of government information and services to the public. The development of a vibrant, indigenous Internet and telecommunications sector is also considered critical for China's long-term global economic competitiveness.[12] Globally, the Internet is rapidly evolving away from personal computers and toward mobile devices, appliances, and vehicles, with the most rapid rate of growth in Internet and mobile-phone use taking place in Africa and the Middle East. The Chinese government's strategy is for Chinese companies to be leaders in mobile Internet innovation, particularly in the developing world. Last year, Premier Wen Jiabao spoke on multiple occasions about the importance of "the Internet of things," encouraging breakthroughs by Chinese companies in what the government has designated as a strategic industry.[13]

Although the government has direct control over websites run by state-operated media as well as its own national- and provincial-level websites, by far the largest portion of the Chinese Internet is run by the private sector (or "government-regulated commercial spaces" according to Min Jiang's taxonomy of Chinese deliberative digital spaces). Chinese networked authoritarianism cannot work without the active cooperation of private companies—regardless of the origin of their financing or where they are headquartered. Every year a group of Chinese Internet executives is chosen to receive the government's "China Internet

Self-Discipline Award" for fostering "harmonious and healthy Internet development."

In Anglo-European legal parlance, the legal mechanism used to implement such a "self-discipline" system is "intermediary liability." It is the mechanism by which Google's Chinese search engine, Google. cn, was required to censor itself until Google redirected its simplified Chinese search engine offshore to Hong Kong. All Internet companies operating within Chinese jurisdiction—domestic or foreign—are held liable for everything appearing on their search engines, blogging platforms, and social-networking services. They are also legally responsible for everything their users discuss or organize through chat clients and messaging services. In this way, the government hands many censorship and surveillance tasks to private companies that face license revocations and forced shutdowns should they fail to comply. Every one of China's large Internet companies has a special department full of employees whose sole job is to police users and censor content.

In 2008, I conducted a comparative study examining how fifteen different Chinese blog-hosting services censored user-created content. The tests revealed that each company used slightly different methods and approaches in its censorship. The specific content censored also varied from service to service. In a number of tests, when I tried to post politically sensitive material such as an article about the parents of students killed in Tiananmen Square, or a recent clash in a remote town in Western China, internal site software would block publication of the post entirely. Other posts could be saved as drafts but were "held for moderation" until a company staffer could make a decision about whether they should be allowed. Other postings simply disappeared within hours of publication.

Lifting the Veil

In June 2010, a report giving Internet users a peek behind the veil of secrecy surrounding corporate complicity in Chinese Internet censorship appeared on the popular Chinese website Sina.com for a few hours before, ironically, being censored. It quoted Chen Tong, the editor of Sina's Twitter-like microblogging service, who described his company's censorship system in some detail: round-the-clock policing; constant coordination between the editorial department and the "monitoring department"; daily meetings to discuss the latest government orders listing new topics and sensitive keywords that must either be monitored or deleted depending on the level of sensitivity; and finally, systems through which both editors and users report problematic content and bring it to the attention of company censors.[14] In April 2009, an employee of Baidu, China's leading search engine, which also runs user-generated content services, leaked a set of detailed documents from Baidu's internal monitoring and censorship department confirming the company's longstanding reputation as an industry leader not only as a search engine and online-services company, but also in censoring both search-engine results and user-generated content. The documents included censorship guidelines; lists of specific topics and words to be censored; guidelines on how to search for information that needs to be deleted, blocked, or banned; and other internal information from November 2008 through March 2009.[15]

In its efforts to manage what the Chinese people can learn, discuss, and organize online, the government deploys a range of other tactics as well. They include:

Cyberattacks: The sophisticated, military-grade cyberattacks launched against Google in late 2009 were targeted specifically at the Gmail accounts of human-rights activists who are either from China or work on China-related issues. Websites run by Chinese exiles, dissidents, and human-rights defenders (most of whom lack the training or resources to protect themselves) have been the victims of increasingly aggressive cyberattacks over the past few years—in some cases, compromising activists' computer networks and e-mail accounts. Domestic and foreign journalists who report on politically sensitive issues and academics whose research includes human-rights problems have also found themselves under aggressive attack in China, with efforts to expose their sources, making it much more risky to work on politically sensitive topics.

Device and network controls: In May 2009, the Ministry of Industry and Information Technology (MIIT) mandated that by July 1 of that year a specific software product called Green Dam Youth Escort was to be preinstalled on all computers sold in China. While Green Dam was ostensibly aimed at protecting children from inappropriate content, researchers outside and within China quickly discovered that it not only censored political and religious content but also logged user activity and sent this information back to a central computer server belonging to the software developer's company. The software had other problems that created opposition to it within U.S. companies. It contained serious programming flaws that increased the user's vulnerability to cyberattack. It also violated the intellectual property rights of a U.S. company's filtering product. Faced with uniform opposition from the U.S. computer industry and strong protests from the U.S. government, the MIIT backed down on the eve of its deadline, making the installation of Green Dam voluntary instead of mandatory.

The defeat of Green Dam, however, did not diminish other efforts to control and track Internet-user behavior at more localized levels— schools, universities, apartment blocks, and citywide Internet Service Providers (ISPs). In September 2009, news reports circulated that local governments were mandating the use of censorship and surveillance products with names such as "Blue Shield" and "Huadun." The purpose of these products appeared similar to Green Dam's, though they involved neither the end user nor foreign companies.[16] Unlike Green Dam, the implementation of these systems has received little attention from foreign media, governments, or human-rights groups.

Domain-name controls: In December 2009, the government-affiliated China Internet Network Information Center (CNNIC) announced that it would no longer allow individuals to register Internet domain names ending in ".cn." Only companies or organizations would be able to use the .cn domain. While authorities explained that this measure was aimed at cleaning up pornography, fraud, and spam, a group of Chinese webmasters protested that it also violated individual rights. Authorities announced that more than 130,000 websites had been shut down in the cleanup. In January 2010, a Chinese newspaper reported that self-employed individuals and freelancers conducting online business had been badly hurt by the measure.[17] In February, CNNIC backtracked somewhat, announcing that individuals would once again be allowed to register .cn domains, but all applicants would have to appear in person to confirm their registration, show a government ID, and submit a photo of themselves with their application. This eliminated the possibility of anonymous domain-name registration under .cn and has made it easier for authorities to warn or intimidate website operators when "objectionable" content appears.

Localized disconnection and restriction: In times of crisis, when the government wants to ensure that people cannot use the Internet or mobile phones to organize protests, connections are shut down entirely or heavily restricted in specific locations. The most extreme case is in the far-northwestern province of Xinjiang, a traditionally Muslim region that borders Pakistan, Kazakhstan, and Afghanistan. After ethnic riots took place in July 2009, the Internet was cut off in the entire province for six months, along with most mobile text messaging and international phone service. No one in Xinjiang could send e-mail or access any website—domestic or foreign. Business people had to travel to the bordering province of Gansu to communicate with customers. Internet access and phone service have since been restored, but with severe limitations on the number of text messages that people can send on their mobile phones per day, no access to overseas websites, and very limited access even to domestic Chinese websites. Xinjiang-based Internet users can only access watered-down versions of official Chinese news and information sites, with many of the functions such as blogging or comments disabled.[18]

Surveillance: Surveillance of Internet and mobile users is conducted in a variety of ways, contributing to an atmosphere of self-censorship. Surveillance enables authorities to warn and harass Internet users either via electronic communications or in person when individuals are deemed to have transgressed certain standards. Detention, arrest, or imprisonment of selected individuals serves as an effective warning to others that they are being watched. Surveillance techniques include:

"Classic" monitoring: While surveillance measures are justified to the public as antiterrorism measures, they are also broadly used to identify and harass or imprison peaceful critics of the regime. Cybercafés—the cheap and popular option for students and the less affluent—are required to monitor users in multiple ways, including identity registration upon entry to the café or upon login, surveillance cameras, and monitoring software installed on computers.

"Law-enforcement compliance": In China, where "crime" is defined broadly to include political dissent, companies with in-country operations and user data stored locally can easily find themselves complicit in the surveillance and jailing of political dissidents. The most notorious example of law-enforcement compliance gone wrong was when Yahoo's local Beijing staff gave Chinese police account information of activist Wang Xiaoning in 2002 and journalist Shi Tao in 2004, leading to their imprisonment. In 2006, Skype partnered with a Chinese company to provide a localized version of its Internet-based phone-calling service, then found itself being used by Chinese authorities to track and log politically sensitive chat sessions by users inside China. Skype had delegated law-enforcement compliance to its local partner without sufficient attention to how the compliance was being carried out.[19]

"Astroturfing" and public outreach: The government increasingly combines censorship and surveillance measures with proactive efforts to steer online conversations. In 2008, the Hong Kong–based researcher David Bandurski determined that at least 280,000 people had been hired at various levels of government to work as "online commentators." Known derisively in the Chinese blogosphere as the "fifty-cent party," these people are paid to write posts that show their employers in a favorable light in online chatrooms, social-networking services, blogs, and comments sections of news websites.[20] Many more people do similar work as volunteers—recruited from the ranks of retired officials as well as college students in the Communist Youth League who aspire to become Party members. This approach is similar to a tactic known as "astroturfing" in U.S. parlance, now commonly used by commercial advertising firms, public-relations companies, and election campaigns around the world in order to simulate grassroots enthusiasm for a product or candidate. In many Chinese provinces, it is now also standard practice for government officials—particularly at the city and county

level—to coopt and influence independent online writers by inviting
them to special conferences and press events.

The central government has also adopted a strategy of using official interactive portals and blogs, which are cited as evidence both
at home and abroad that China is liberalizing. In September 2010, the
CCP launched an online bulletin board called "Direct to Zhongnan-
hai," through which the public was invited to send messages to China's
top leaders. Since 2008, President Hu Jintao and Premier Wen Jiabao
have held annual "web chats" with China's "netizens." An official
"E-Parliament" website, on which citizens are invited to post policy
suggestions to the National People's Congress, was launched in 2009.
The 2010 official government white paper lists a variety of ways in
which the Chinese government solicits public feedback through the
Internet. It states: "According to a sample survey, over 60 percent of
netizens have a positive opinion of the fact that the government gives
wide scope to the Internet's role in supervision, and consider it a
manifestation of China's socialist democracy and progress."[21]

All of this is taking place in the context of the Chinese government's
broader policies on information and news control. In December 2009,
the Committee to Protect Journalists listed China as the world's worst
jailer of journalists. In recent testimony before the U.S. Congress, Joshua Rosenzweig of the Dui Hua Foundation presented an array of statistics to support a grim conclusion:

> Over the past two-and-a-half years in particular, roughly since the beginning
> of 2008, there has been a palpable sense that earlier progress towards rule of
> law in China has stalled, or even suffered a reversal, and there is mounting
> evidence that a crackdown is underway, one particularly targeting members
> of ethnic minorities, government critics, and rights defenders.[22]

Thus online public discourse is indeed expanding—with government
encouragement. The government is creating and promoting the impression both at home and abroad that China is moving in the direction of
greater democracy. At the same time, the Chinese people's ability to
engage in serious political dissent or to organize political movements
that might effectively challenge the CCP's legitimacy has actually diminished, and the consequences for attempting such activities are more
dire than they were ten years ago.

Networked Authoritarianism Beyond China

In their most recent book surveying Internet censorship and control around the world, Ron Deibert and Rafal Rohozinski warn that
"the center of gravity of practices aimed at managing cyberspace has
shifted subtly from policies and practices aimed at denying access to
content to methods that seek to normalize control and the exercise

of power in cyberspace through a variety of means." This article has described a range of ways in which China is near the forefront of this trend. Deibert and Rohozinski divide the techniques used by governments for Internet censorship and control into three "generations": The "first generation" of techniques focuses on "Chinese-style" Internet filtering and Internet-café surveillance. "Second-generation" techniques include the construction of a legal environment legitimizing information control, authorities' informal requests to companies for removal of information, technical shutdowns of websites, and computer-network attacks. "Third-generation" techniques include warrantless surveillance, the creation of "national cyberzones," state-sponsored information campaigns, and direct physical action to silence individuals or groups.[23]

While Deibert and Rohozinski characterize Chinese cybercontrols as being largely first generation, the Chinese government aggressively uses all the second- and third-generation techniques and has been doing so for quite some time. Indeed, the second- and third-generation techniques are essential because the Great Firewall alone is ineffective and permeable.

Deibert and Rohozinski point out that a number of governments, particularly those in Russia and several former Soviet republics, have bypassed the first-generation controls almost completely and instead are concentrating their energies on second- and third-generation controls, most of which (with the jarring exception of "direct physical action to silence individuals or groups") are more subtle, more difficult to detect, and more compatible with democratic or pseudodemocratic institutions. The Russian-language Internet, known by its denizens as "RUNET," is thus on the cutting edge of techniques aimed to control online speech with little or no direct filtering.[24]

Research in the Middle East and North Africa shows that while Internet filtering is more common and pervasive throughout that region, governments are increasing the use of second- and third-generation techniques. Many governments in the region have cracked down on online dissent through the skillful use of family-safety measures and antiterrorism laws. At the same time, they have made substantial investments in Internet and telecommunications infrastructure, recognizing that connectivity is essential for economic success.[25]

Some second- and third-generation controls are also used by democratically elected governments, including those of South Korea and India.[26] Intermediary censorship is deployed in a range of political systems to silence antiregime speech, fight crime, or protect children. The concept of holding service providers liable has become increasingly popular among lawmakers around the world, including in Western Europe—where the main goals are to combat intellectual-property theft and protect children. In the United States, activists are concerned about the weakening of due process, which has allowed government access to networks owned and run by corporations, all in the name of combating cybercrime and cyberwarfare. Even the Chinese government has adopted a very similar language of cybersecurity to justify its Internet-control structures and procedures. Deibert and Rohozinski are right to warn that "many of the legal mechanisms that legitimate control over cyberspace, and its militarization, are led by the advanced democratic countries of Europe and North America."[27]

Chinese authoritarianism has adapted to the Internet Age not merely through the deployment of Internet filtering, but also through the skilled use of second- and third-generation controls. China's brand of networked authoritarianism serves as a model for other regimes, such as the one in Iran, that seek to maintain power and legitimacy in the Internet Age. In Russia and elsewhere there is a further, disturbing trend: Strong governments in weak or new democracies are using second- and third-generation Internet controls in ways that contribute to the erosion of democracy and slippage back toward authoritarianism. This situation is enabled by a weak rule of law, lack of an independent judiciary, weak guarantees for freedom of speech and other human-rights protections, heavy or untransparent regulation of industry (particularly the telecommunications sector), and weak political opposition that is rendered even weaker by clever manipulation of the media, legal system, and commercial-regulatory system.

It is clear that simply helping activists to circumvent first-generation censorship and training them in the use of new technologies for digital activism without also addressing the second- and third-generation controls deployed by their governments is insufficient, sometimes counterproductive, and potentially dangerous for the individuals involved. Weak rule of law and lack of accountability and transparency in the regulation of privately owned and operated Internet platforms and telecommunications networks facilitate the use of second- and third-generation controls, which pose a great threat to activists. Therefore, strong advocacy work at the policy and legislative level aimed at improving rule of law, transparency, and accountability—in government as well as the private sector—is more important than ever.

The business and regulatory environment for telecommunications and Internet services must become a new and important focus of human-rights activism and policy. Free and democratic political discourse requires Internet and telecommunications regulation and policy making that are transparent, accountable, and open to reform both through independent courts and the political system. Without such baseline conditions, opposition, dissent, and reform movements will face an increasingly uphill battle against progressively more innovative forms of censorship and surveillance.

NOTES

1. Hillary Rodham Clinton, "Remarks on Internet Freedom," Washington, D.C., 21 January 2010; available at *www.state.gov/secretary/rm/2010/01/135519.htm.*

2. "Chinese State Security Arrests, Indictments Doubled in 2008," *Dui Hua Human Rights Journal,* 25 March 2009; available at *www.duihua.org/hrjournal/2009/03/chinese-state-security-arrests.html.*

3. Larry Diamond, "Liberation Technology," *Journal of Democracy* 21 (July 2010): 82.

4. Lokman Tsui, "The Great Firewall as Iron Curtain 2.0: The Implications of China's Internet Most Dominant Metaphor for U.S. Foreign Policy," paper presented at the sixth annual Chinese Internet Research Conference, Hong Kong University, 13–14 June 2008; available at *http://jmsc.hku.hk/blogs/circ/files/2008/06/tsui_lokman.pdf.*

5. Guobin Yang, *The Power of the Internet in China: Citizen Activism Online* (New York: Columbia University Press, 2009), 213.

6. Jack Linchuan Qiu, *Working-Class Network Society: Communication Technology and the Information Have-Less in Urban China* (Cambridge: MIT Press, 2009), 243.

7. Yongnian Zheng, *Technological Empowerment: The Internet, State, and Society in China* (Stanford: Stanford University Press, 2008), 164–65.

8. Nicholas D. Kristof, "Death by a Thousand Blogs," *New York Times,* 24 May 2005; available at *www.nytimes.com/2005/05/24/opinion/24kristoff.html.*

9. Baogang He and Mark Warren, "Authoritarian Deliberation: The Deliberative Turn in Chinese Political Development," paper presented at the Annual Meeting of the American Political Science Association, Boston, 28–31 August 2008; forthcoming, *Perspectives on Politics,* June 2011.

10. Min Jiang, "Authoritarian Deliberation on Chinese Internet," *Electronic Journal of Communication* 20 (2010); available at *http://papers.ssrn.com/sol3/papers.cfm?abstract_id=1439354.*

11. Rebecca MacKinnon, "Chinese Internet Research Conference: Getting Beyond 'Iron Curtain 2.0,'" *RConversation,* 18 June 2008; available at *http://rconversation.blogs.com/rconversation/2008/06/chinese-inter-1.html.*

12. "The Internet in China," Information Office of the State Council of the People's Republic of China (SCIO), 8 June 2010; available at *http://china.org.cn/government/whitepaper/node_7093508.htm.*

13. Robert McManus, "Chinese Premier Talks Up Internet of Things," *ReadWriteWeb,* 19 January 2010; available at *www.readwriteweb.com/archives/chinese_premier_internet_of_things.php.*

14. Jonathan Ansfield, "China Tests New Controls on Twitter-Style Services," *New York Times,* 16 July 2010; available at *www.nytimes.com/2010/07/17/world/asia/17beijing.html.* The full Chinese-language text of the report (which was deleted by censors from the original source) was reproduced by Radio France Internationale at *www.chinese.rfi.fr.*

15. Xiao Qiang, "Baidu's Internal Monitoring and Censorship Document Leaked," *China Digital Times,* 30 April 2009; available at *http://chinadigitaltimes.net/2009/04/baidus-internal-monitoring-and-censorship-document-leaked/.*

16. Owen Fletcher, "China Clamps Down on Internet Ahead of 60th Anniversary," IDG News Service, 25 September 2009; available at *www.pcworld.com/article/172627/ china_clamps_down_on_internet_ahead_of_60th_anniversary.html*; and Oiwan Lam, "China: Blue Dam Activated," *Global Voices Advocacy,* 13 September 2009; available at *http://advocacy.globalvoicesonline.org/2009/09/13/china-blue-dam-activated.*

17. Oiwan Lam, "China: More than 100 Thousand Websites Shut Down," *Global Voices Advocacy,* 3 February 2010; available at *http://advocacy.globalvoicesonline. org/2010/02/03/china-more-than-100-thousand-websites-shut-down.*

18. Josh Karamay, "Blogger Describes Xinjiang as an 'Internet Prison,'" BBC News, 3 February 2010; available at *http://news.bbc.co.uk/2/hi/asia-pacific/8492224.stm.*

19. Nart Villeneuve, "Breaching Trust: An Analysis of Surveillance and Security Practices on China's TOM-Skype Platform," Open Net Initiative and Information Warfare Monitor, October 2008; available at: *www.nartv.org/mirror/breachingtrust.pdf.*

20. David Bandurski, "China's Guerilla War for the Web," *Far Eastern Economic Review,* July 2008.

21. SCIO, "The Internet in China."

22. Joshua Rosenzweig, "Political Prisoners in China: Trends and Implications for U.S. Policy," Testimony to the Congressional-Executive Committee on China, 3 August 2010; available at *www.cecc.gov/pages/hearings/2010/20100803/statement5.php.*

23. Ronald Deibert and Rafal Rohozinski, "Control and Subversion in Russian Cyberspace," in Ronald Deibert et al., eds., *Access Controlled: The Shaping of Power, Rights, and Rule in Cyberspace* (Cambridge: MIT Press, 2010), 23.

24. Deibert and Rohozinski, "Control and Subversion in Russian Cyberspace," in *Access Controlled,* 15–34.

25. "MENA Overview," *Access Controlled,* 523–35.

26. Michael Fitzpatrick, "South Korea Wants to Gag the Noisy Internet Rabble," *Guardian.co.uk,* 9 October 2008; available at *www.guardian.co.uk/technology/2008/ oct/09/news.internet*; and John Ribeiro, "India's New IT Law Increases Surveillance Powers," IDG News Service, 27 October 2009; available at *www.networkworld.com/ news/2009/102709-indias-new-it-law-increases.html.*

27. Deibert and Rohozinski, "Beyond Denial: Introducing Next-Generation Information Access Controls," 6.

III

Liberation Technology
in the Middle East

7

USHAHIDI AS A LIBERATION TECHNOLOGY

Patrick Meier

Patrick Meier *serves as Director of Crisis Mapping at Ushahidi and previously codirected Harvard's Program on Crisis Mapping and Early Warning. He also consults extensively for international organizations in Africa, Asia, and Europe. Meier holds a doctorate from the Fletcher School and was a predoctoral fellow at Stanford University's Center on Democracy, Development, and the Rule of Law.*

One of the most dramatic and recent innovations in liberation technology has been the combination of "geomapping" with other new communications technologies, allowing citizens to document crises, corruption, and abuses around the world. In this chapter, I examine one such geomapping liberation technology, Ushahidi, and discuss its role in Egypt's 2010 parliamentary elections.

Ushahidi (meaning "testimony" in Swahili) is a simple web-based mapping platform originally designed to map reports of human-rights abuses using text messages (SMS), e-mails, and information submitted via an online form. The first Ushahidi map was created to show Kenya's postelection violence in January 2008. Bloggers in Kenya and from the Kenyan diaspora launched the platform to map human-rights violations that would otherwise have gone completely undocumented by the mainstream media and official election observers. Ushahidi Inc., a nonprofit technology company, was created several months after the elections to improve the mapping platform and make it free and open-source and thus widely usable. Several subsequent versions of the platform have since been used to create more than ten-thousand live maps in more than 140 countries. Those engaged in these mapping projects have included humanitarian and human-rights organizations, media companies, civil society groups, political and environmental activists, and distributed volunteer networks. The Ushahidi platform saw particular use as a "liberation technology" in Egypt during the country's parliamentary elections in the fall of 2010.

Why focus on Ushahidi? The platform represents an important con-
vergence of new technologies.[1] SMS, Twitter, Facebook, YouTube,
Flickr, smartphone apps, voicemail, and email can all be combined with
Ushahidi. Examining Ushahidi instead of studying the impact of certain
tweets or YouTube videos in isolation allows for a focus on broader
multimedia content. Focusing on the Ushahidi platform also facilitates
the study of concrete uses of social media, such as election monitoring.
Elsewhere in this volume, Larry Diamond has referred to the Ushahidi
platform as an example of a liberation and accountability technology.[2]
What is missing, however, is research to support these claims.

Using ICTs in Nonpermissive Environments

Information and Communication Technologies (ICTs) are used in
various ways and with varying degrees of success to promote democracy, development, rule of law, and transparency in nonpermissive environments. The 2011 revolutions in North Africa and the Middle East
represent the most striking recent uses of ICTs to drive political change
in repressive environments. But ICTs have also been used to monitor
and address issues related to corruption, unemployment, elections, public health, and local governance in dozens of countries around the world.

Mobile-communication technology has been the most rapidly adopted technology in history, far outpacing adoption rates of other technologies, even the Internet.[3] The latest statistics from the International Telecommunications Union counted more than five-billion mobile phones
worldwide at the end of 2010, a 25 percent increase over just the previous year. Half a billion people worldwide now access the Internet by
mobile phone, and it is estimated that this number will double by 2015.
By the end of the decade, some expect the number of mobile wireless
devices to pass 50 billion, a staggering tenfold increase. The technolo-
gy-adoption statistics for Africa are equally astounding, with users of
mobile-communication technology soaring from 2 percent to nearly 30
percent of the population in the first decade of this century.

New ICTs are radically different from traditional communication
tools. Radio and television have been owned and controlled by the very
few, and the telephone and telegraph did not have broadcasting capabilities. By contrast, the majority of content created and shared online
today is generated by a vast multiplicity of users, allowing the many to
converse with the many without undue centralization and control. The
scaling of these conversations is where some believe their power lies,
since access to conversations is more important politically than access
to information.[4]

The use of ICTs figured prominently during the 2011 revolutions in
Tunisia and Egypt. As one Egyptian activist tweeted, "We use Facebook to schedule our protests, Twitter to coordinate, and YouTube to

tell the world."[5] These social-media tools are not simply used in the same way as older technologies to coordinate and document unfolding events; instead, they are increasingly used to create live maps. Dozens of live maps were created to monitor the events in Tunisia, Egypt, Syria, and Yemen, for example. Indeed, we are starting to witness the rise of a "mapping reflex." As one Russian blogger puts it: "If radio gave each event a sound, TV an image, then this relatively new 'mapping reflex' gave each event a geographic location."[6] In many ways, this mapping reflex resembles the "Wikipedia effect"—the creation and real-time editing of Wikipedia entries to document live breaking news. In the case of the Ushahidi platform, contributors edit a map instead of a wiki.

Social media also played a pivotal role during the Libya crisis in 2011. Indeed, Luis Moreno-Ocampo, the International Criminal Court (ICC) prosecutor, cited Facebook and other social media as a key influence on the ICC's decision to take action in Libya.[7] In addition, the UN's Office for the Coordination of Humanitarian Affairs (OCHA) launched a live social-media map of Libya (using the Ushahidi platform) to better inform its humanitarian-relief operations.[8] This map was largely based on reports that were crowdsourced (gathered from the public) through social-media sources like Twitter and YouTube.

Official election-monitoring organizations are also looking to ICTs to improve the speed and reliability of their efforts, particularly in contested elections. The National Democratic Institute (NDI) increasingly uses SMS to monitor elections. Indeed, as Ian Schuler, senior manager of ICT programs at NDI, writes, "the speed of communication and processing, the flexibility, and the coverage SMS can provide give monitoring organizations a powerful tool for organizing volunteers and responding instantly to an evolving election environment."[9]

Ushahidi Technology and Its Applications

In the wake of Kenya's controversial national elections on 27 December 2007—which were widely seen as riddled with fraud—deadly riots erupted, claiming eight-hundred lives within a few weeks. President Mwai Kibaki downplayed the scope of the violence and placed restrictions on the national media's coverage of it. At the same time, international election-monitoring organizations refused to share the data they had collected. The full extent of the violence was largely unknown.

Against this backdrop, Kenya's most prominent blogger, Ory Okolloh, began blogging extensively about the violence. Having a wide readership, she received continuous streams of information from her readers, who were documenting numerous human-rights violations across the country. Okolloh was soon overwhelmed with the volume of information and could not keep blogging fast enough. She was later forced to flee to South Africa following a number of death threats. As Okolloh continued blogging from Johannesburg, she suggested in a blog post that a Google map "mashup" be set up to allow others to document human-rights violations directly, since she could not keep up.[10] Fellow Africa and technology bloggers Erik Hersman, David Kobia, and Juliana Rotich read the post and decided to act on her suggestion. Thus was born Ushahidi.

The Ushahidi platform is a free and open-source mapping software that allows anyone to create a live and rich multimedia map of an event or situation. Unlike the standard Google Maps, the Ushahidi platform allowed witnesses in Kenya to text in their reports of human-rights violations using SMS. A simple SMS "short code" was set up with the telecommunications company Safaricom to make this happen. The bloggers shared the map on their blogs to get the word out and thus began to crowdsource crisis information. They documented information on human-rights violations that would otherwise have gone largely unnoticed by mainstream media and election-monitoring organizations.

The Ushahidi software has gone through several important upgrades over the past three years. The platform can now be integrated with SMS, Twitter, e-mail, voicemail, Facebook, and soon Flickr and YouTube. Ushahidi Inc. has also developed dedicated smartphone apps for the platform. These are freely available for the iPhone, Android, and other java-enabled phones. In 2010, the company released Ushahidi 2.0, which allows third-party developers to develop customized apps or plug-ins for the core platform, thereby extending the platform's capabilities considerably. This latest version of Ushahidi makes it easier to map multimedia content such as photographic evidence and video footage. In addition, Ushahidi recently launched Crowdmap, a hosted version of the Ushahidi platform that further reduces the barriers to creating live maps.

The Ushahidi platform has been deployed in over forty countries for a wide range of uses, including election observation, human-rights monitoring, disaster response, civil resistance, and environmental-impact reporting. For example, civil society groups have used the platform to observe elections in Afghanistan, Burundi, Egypt, India, Kenya, Kyrgyzstan, Lebanon, the Philippines, Mexico, Mozambique, Sudan, and Tanzania. Ushahidi has also been used for disaster response and human-rights monitoring in Australia, Chile, the Czech Republic, Haiti, Pakistan, Poland, South Africa, Tunisia, and the United States, and to map the protests that unfolded during the recent revolutions in North Africa.

The impact of Ushahidi platforms, however, has been unclear. Very few evaluations—let alone rigorous ones—have been carried out. Several groups that have used the Ushahidi platform claim that its impact is obvious. But the evidence remains largely anecdotal and the analysis rather thin.

Egypt was selected as a country case study for this chapter for three reasons. First, the Ushahidi platform has been used multiple times in

Egypt since the 2010 parliamentary elections. Thus, insights can also be drawn from more recent examples, such as the protests during the 2011 revolutions in North Africa. Second, Egypt under Hosni Mubarak qualified as a repressive state—an important criterion given this book's focus on liberation technology in authoritarian contexts. Third, the Egyptian group that has repeatedly deployed the Ushahidi platform is continuing to do so, not only in Egypt but now also in Lebanon, Syria, Tunisia, and Yemen. Hence, interested scholars and practitioners will be able to draw on the findings from this chapter to inform future research on and applications of the technology in Egypt and beyond.

The U-Shahid (*shahid* means "witness" in Arabic) project in Egypt—run by the Development and Institutionalization Support Center (DISC), an Egyptian organization based in Cairo—first used Ushahidi during parliamentary elections in November and December 2010. Just days after Internet access was restored during the civil-resistance movement in early 2011, DISC used the platform again to map the protests against the Mubarak regime and its human-rights abuses. The Egyptian group has since launched a Ushahidi platform to map feedback on constitutional amendments, and it plans to use Ushahidi to map corruption as well. A separate Cairo-based group also began using the Ushahidi platform in 2010 for a project called Harassmap, which enables Egyptian women to report cases of harassment and increase the visibility of this chronic problem.

The Use of Ushahidi in Egypt

Until recently, Egyptians had only been able to approve or reject a presidential candidate appointed by the parliament, which was dominated by Mubarak's National Democratic Party (NDP).[11] Not surprisingly, the media landscape in Egypt was mostly controlled by the establishment during this time. A constitutional amendment approved in a 2005 referendum paved the way for multiparty presidential elections, and Egyptian youth became more and more interested in having a voice and an active part in the political discourse in their country.

The Ministry of Interior was well aware of these changes in political activism, especially with respect to the use of social networks. It took steps to level the social media battlefield by adopting a strategy similar to that of digital activists. On 1 July 2010, the ministry reportedly established a special department of fifteen individuals who took shifts in order to operate 24 hours a day.[12] Their main task was to monitor Facebook content such as groups, pages, and chats and to publish reports countering online criticism of President Mubarak and his son Gamal. In addition, the NDP recruited groups of young people to create Facebook pages and groups to support the president, his son, and the ruling party.

In this political and social-media context, DISC used the Ushahidi platform to launch their U-Shahid project. The head of DISC, Kamal Nabil, had first come across Ushahidi during training in Washington, D.C., organized by Freedom House in early 2010. The goal of the U-Shahid project was to monitor the parliamentary elections in late 2010. This independent initiative became particularly important when the Mubarak regime announced that it would not permit any official international election-monitoring groups into the country. Despite the pessimism and despair about the political situation in Egypt, "the undercurrent of digital activism was tangible," according to a foreign activist who joined DISC. The use of social media and Facebook in particular increasingly enabled the youth to engage in a political context in which the physical elimination of the opposition was the norm. Blogs and Facebook groups filled the vacuum created by the lack of a real political debate in Egypt, and they increasingly emerged as an alternative political scene where a discourse on democracy and human rights was still possible.

The U-Shahid project was rather simple on paper—use the Ushahidi platform to monitor the elections by allowing people to send SMS, tweets, Facebook comments, voicemail, e-mail, and reports via webform to the live map. DISC decided to draw on both crowdsourced reporting and "blogger-sourced" information. This meant navigating the restrictions imposed by Egyptian national-security officials, while getting the word out to the wider public and training a large network of 130 trusted bloggers across the country. Despite the restrictions, training for the latter took place in five major cities: Cairo, Alexandria, Assiut, Mansoura, and Port Said. DISC translated its Ushahidi platform entirely into Arabic, since the U-Shahid project was meant to be "an Egyptian project for Egyptians," as one blogger stated. Egyptian software developers integrated Twitter, Flickr, and YouTube with Ushahidi. Since Facebook was and continues to be an important platform for Egyptian youths, the group also created a feature to enable comments on a Facebook wall to be easily mapped on the Ushahidi platform.

DISC formulated clear goals for U-Shahid. The first was to help Egyptian citizens and international observers learn more about the electoral process. Second, the project aimed to report and seek redress when electoral laws were violated. Third, DISC sought to raise awareness about citizen rights and the importance of participation in the electoral process. Finally, DISC wanted to use the U-Shahid project to empower local partners to advocate for closer adherence to electoral laws and fair practices during both the campaign and election period.

How did the team do? During the elections, DISC sources mapped 2,700 reports, which included 211 supporting pictures and 323 videos. The team of Egyptian bloggers was also able to verify more than 90 percent of the content that ended up on the map by using basic journalistic techniques such as triangulation and follow-up. Most of the "crowdsourced" information, however, came from the preestablished network of trusted bloggers, and thus did not require immediate verification. In total, the web-based map received over 40,000 hits, the vast majority of which came from within Egypt. (Interestingly, the next highest number originated from Saudi Arabia, with just under 5,000 hits.) The group proactively disseminated this information, through both new and traditional media. Their efforts were featured on Egyptian television, on BBC Arabic, and in dozens of articles in ten different languages.

Even before it was formally launched, the project got the attention of the Egyptian government. An Egyptian state security official contacted Kamal Nabil and told him that his name was recurring "too often" in phone conversations between activists. The Ministry of Interior subsequently shadowed the project in different ways: by tapping the cell phones of the core team of bloggers; by requesting copies of the agendas for all meetings related to U-Shahid; and by requiring that a list of all individuals trained in the use of the platform be submitted to them. Email addresses, Facebook pages, and Twitter accounts of the core team were reportedly all under surveillance from the start of the project, and the Ministry of Interior openly asked Nabil what his reaction would be if they were to shut down the U-Shahid project before the elections.

Moreover, several new Facebook groups were launched to attack the core team by accusing its members of being affiliated with the United States, on the grounds that they had participated in the Freedom House–organized conference in Washington, D.C., earlier that year. Some of those Facebook groups called on young Egyptians to "watch out" for projects that could endanger the political independence and integrity of the country. Activists reacted to these attacks by waging a virtual battle. Once a government-supported group was identified, dozens of activists would write on the group's wall, basically occupying the entire wall with contrary opinions.

DISC was well aware that technology alone would not change the political situation in Egypt. They also knew that Egyptian state security could shut down the project and block access to the website whenever it wanted. Furthermore, everyone involved in the project knew full well that their involvement in U-Shahid could get them arrested. At the end of a training workshop in Cairo, one participant told a trainer, "You know, we may all end up in jail, but before this I thought there was no hope to change anything. Now I can even dare to think it is worth a try."

The impact of the U-Shahid project on the political space in Egypt is difficult to assess. According to the lead trainer of the project, more than 1,500 election complaints were officially submitted to the courts. It is unclear, however, whether any of these came from or were influenced by the content mapped on the Ushahidi platform. Even overlap between U-Shahid's 2,700 reports and the courts' 1,500 would highlight the value of the project, since the latter's data could be used to triangulate or bolster evidence submitted to the courts. Unfortunately, accessing the complaints received by the court has not been possible. Instead, the reports submitted to the Ushahidi platform during the parliamentary elections and civil resistance were analyzed to assess the potential impact of the project and identify trends or recurring patterns in the reports.

The topics most frequently addressed in reports submitted to the Ushahidi platform included bribes for votes, police closure of roads leading to polling centers, the destruction and falsification of election ballots, evidence of violence in specific locations, the closure of polling centers before the official time, and the banning of local election observers from polling centers. Perhaps most striking about the reports, however, is how specific they were in terms of location and details. For example, reports that documented the buying of votes often included the amount paid for the vote. Other reports documented nonfinancial bribes, including mobile phones, food, gas, and even "sex stimulators," "Viagra," and "Tramadol tablets."

Additional incidents mapped on the Ushahidi platform included reports of deliberate power cuts to prevent people from voting. One voter complained that "in Al Saaida Zaniab election center we could not find my name in voters' lists, [even though] I voted in the same committee. Nobody helped to find my name on [the] list because the electricity cut out." Voters also complained about the lack of phosphoric ink for voting and the fact that they were not asked for their IDs to vote. Reports also documented harassment and violence by thugs, often against Muslim Brotherhood candidates, and the busing in of people from the ruling party. For example, one report noted that "Oil Minister Samih Fahmi, who is a national nominee for Peoples Council in Al Nassr City, used his power to mobilize employees to vote for him. The employees used the company's buses, which carried the nominee's pictures, to go to the election centers." Several hundred reports included pictures and videos, some clearly documenting obvious election fraud. There were also some reports, however, that documented calm (e.g. "everything is ok") at certain voting centers.

The evidence documented by the U-Shahid team had the potential to create greater political accountability by shining more light on the process of election fraud. It is doubtful, however, that the U-Shahid project actually deterred fraud. The project was simply not operating at a scale of visibility sufficient to change behavior. Documenting 2,700 instances of election irregularities is impressive given the many challenges of operating in a repressive environment and the fact that this was the first use of the Ushahidi platform in Egypt. But it might have required 270,000 reports documenting all facets of the election—before, during, and after—with tens of thousands of original videos and photographic evidence to deter those planning to commit fraud. While the

2,700 reports mapped on U-Shahid came from more than 100 individuals, this too was insufficient to have a large-scale and long-term impact. If 100,000 people or more had participated in sending in reports, this might have reached the scale at which the U-Shahid project could have had more meaningful impact.

What the Activists Say

According to members of the U-Shahid project, the use of Ushahidi increased civic participation in election observation, primarily because the web-based nature of the platform allowed for ideas to be more easily expressed online. The Ushahidi platform provided an easy and public way for ordinary Egyptians to share what they were witnessing—fraud, violence, and the like. One activist noted that the technology allowed more people to "make small, low-risk contributions, like sending SMS or an email."

The lead trainer for the project explained that, in the past, "NGOs had been more visibly involved in election monitoring, which made it more dangerous, and observers had to be accredited by formal organizations. But with Ushahidi, anyone could report, even if they had never been observers before. They didn't have to register." In addition, the training sessions did not require technological savvy and instead were focused on political conversations and participation.

When asked why the regime had not shut down the Ushahidi platform, one blogger explained that "many of the activists who began using Ushahidi had many followers on Facebook and Twitter; they also had the attention of the international media, which could draw unwanted attention to the regime's actions. They were connected with people in the U.S. Congress, directors of international human-rights NGOs, and so on." This explanation is in line with Philip Howard's finding that "having an online civil society is a key ingredient of the causal recipe for democratization."[13]

Interviewees also noted that the Mubarak regime had a limited understanding of technology, and therefore the relationship between the state and DISC did not necessarily become more contentious over time. As one key person at DISC noted, "They [government officials] didn't quite understand the technology and were afraid of the Ushahidi platform." Another activist added that "the government was nervous; they didn't feel in control. And the government is usually behind anyways, they're not in the driver's seat [when it comes to technology]." Another reason why the relationship did not become more contentious is because DISC remained fully transparent about the project. "We stressed the technical aspect of the project, and remained fully open and transparent about our work. We gave Egyptian National Security a dedicated username and password [to access the Ushahidi platform], one that we could control and [thus use to] monitor [their actions]. This gave them a false sense of control; we could restore anything they deleted."

In terms of organizational issues, the team was able to leverage existing networks of activists and remain flexible. As noted in one interview, the Egyptian state's hierarchical organization made it less effective in responding quickly to a changing situation, while activists could do so almost in real time since their lines of command were far more diffuse than the government's. One activist remarked that government officials "don't understand how we work; we can learn very fast, but the government has many rules and processes. They have to write up reports, submit them for approval, and allocate funding to acquire technology. But for us, we don't need permission. If we want to use Tor, we simply use Tor." As another activist explained:

> The government had two mechanisms at its disposal to get in our way: intimidation and bribes. But to influence these two mechanisms, you have to access the leadership, and with technology, this connection is a lot harder to make; it becomes more about distributed [rather than hierarchical] leadership. The government couldn't just target one person [i.e., the director of DISC] to shut down the project—they had to target 100. This gave a sense of empowerment to the people.

When asked whether the Ushahidi platform led to more or less access to the political system in Egypt, all interviewees answered "more access." One activist explained that members of the U-Shahid project "were some of the most interviewed people on TV, [which] gave us access to the government and the public; we also had a lot more access to more candidates who wanted to have their representatives trained on the Ushahidi platform . . . and were also invited to train journalists. . . . We also got access to other international organizations that promoted our initiative." Another activist argued that the use of the Ushahidi platform "created more transparency around the elections, allowing easier access than in any previous election." When asked whether any of the 2,700 reports submitted to the Ushahidi platform had made their way to the courts, however, activists replied that it was difficult to know for sure.

According to some participating bloggers, there was a sense that doing anything more than resorting to online tools would lead to physical harm. While activists may have felt safer organizing online than in person, they did face some "opportunity costs" in using the Ushahidi platform. "We were afraid that the government would be filtering reports coming to us and that they would track the reports back to the people who sent them," one activist noted. Another added that this fear might have dissuaded some people from submitting evidence. The lead trainer said, "Yes, definitely, we faced some serious constraints. For example, very few people sent in reports via SMS—at most one percent of the reports we received. One reason for this was that everyone knew that the government could track and control SMS."

In addition, the "timely compilation of reports made a huge difference. In the past, covering elections would mean the media giving quick superficial updates, or established organizations giving a comprehensive bigger picture, but only much later. With Ushahidi, you have the big picture immediately." As the lead trainer for the project noted, "We had never seen so many videos on YouTube about the elections. It was simply the right time [to do a project like U-Shahid]. . . . The Ushahidi platform definitely helped contribute to this significant increase [in user-generated content around the elections]."

On a related note, an activist explained that the U-Shahid project was able to

> cover a lot more information than the traditional media; while they had their own coverage, we provided more timely information, which is very important for the media. We gave them evidence: pictures, videos, and statistics. The media doesn't have access to all this kind of information, so the reports on the Ushahidi platform were a treasure for them. Even if the government was trying to pressure the media, the information was too valuable for them not to show it.

In a way, the information displayed on the Ushahidi platform not only circumvented the state media, but also coopted some national media outlets.

Finally, the launch of U-Shahid inspired some "copycats," as four additional Ushahidi platforms sprang up shortly before the elections, including one launched by the Muslim Brotherhood to document harassment of their candidates. This proliferation of Ushahidi platforms helped to frame an alternative discourse during the election period.

All the interviewees stated that the regime was not particularly effective in using technology to foster patriotism. "If they had been, they would have stopped the revolution," one blogger noted. That said, one activist remarked that the government did try:

> They had an army of bloggers who would go to activist websites to lobby them and to report them so they would have their Facebook pages suspended. They also tried to do that with some websites, but we had a secure system. There were attempts by the government to overload our website with many fake reports . . . but we were on it and we were able to delete them. This happened for a minute or two every three hours or so—attacks, overload—but eventually they gave up.

Egyptian activists believe that the Ushahidi technology is notably different from ICTs that they have used to organize and mobilize in the past. One activist recalled that an election-monitoring NGO had used a map to monitor previous elections, but the resulting website had a page rank of six million, even though the NGO had paid staff thousands of dollars to create the web-based map. "The map was not easy to use or to browse," the activist said. "The people behind the map were professionals at election monitoring, but they were not professionals in technology." In contrast, the Ushahidi map for the U-Shahid project "had a 40,000 ranking worldwide. Plus it was open source and reached tens of thousands of people."

Regarding the cat-and-mouse game between the state and the civil society movements, one activist said,

> We did a lot of scenario building, considered many 'what if' situations. The fact that we were so well prepared is why [the regime] could not touch us. We tried to connect all the data on Facebook and Twitter so that if they closed our Ushahidi map, we would move to a new domain name and let all our followers know. We also had a large database of SMS numbers, which would allow us to text our followers with information on the new website.

Another blogger noted that "because we were well prepared, we knew they could not arrest all of us on the day of the election, and just in case, we trained a group in Lebanon who could take over all operations if we were stopped." The team also set up a phone tree in case of arrest and made multiple copies of the platform.

Another key activist observed:

> Technology by nature is a very neutral tool. But the most important thing is information. Information is the key that drives political discourse and media debates. Information wants to be found. Those who want to suppress it will have a harder time. So people in favor of spreading information are going to win.

The lead trainer of the project opined that "regardless of technology, numbers still matter, and there will always be more citizens than politicians. So I believe in the power of numbers and organization."

Was U-Shahid Successful?

The activists behind U-Shahid set out to achieve concrete goals: to inform Egyptians and international observers about the electoral process, to expose and remedy election violations, to raise awareness about citizens' rights and the importance of electoral participation, and to empower local partners to advocate for fair and clean election practices.

Did they have the impact that they intended? Largely yes, according to those interviewed. They were able to publish and widely disseminate the electoral laws of Egypt, the Egyptian constitution, applicable human-rights conventions, and up-to-date news on the electoral process and campaign. They received some 40,000 hits on their dedicated map, leveraged the web through Facebook, Twitter, and blogs, and received a notable amount of national and international media coverage. They mapped 2,700 reports, with more than 90 percent of them verified. But the project also fell short of achieving some of its goals. In terms of the 1,500 cases of reported electoral violations submitted to the Egyptian courts, the lead trainer for the project noted that they "don't know if those violation complaints are related to the use of this [U-Shahid] platform, or what impact the platform has or will have, in any of those cases." In addition, the group was unable to involve certain sectors of society in the project and to overcome all the technical and political barriers. Finally, the lead trainer stated that "We weren't able to set measurable outcomes for the impact of the project in terms of change . . . but we have time to get better."

The U-Shahid project had some impact on the political space and discourse in Egypt. The use of free and open-source technology meant that DISC faced lower costs, while the use of Facebook, Twitter, and other social-networking platforms also helped to shape a sense of collective identity (although this community largely existed before the elections). A leading Egyptian activist remarked that thanks to its free and open-source technology, as well as its distributed, user-generated approach, the U-Shahid project was less costly than traditional election monitoring. The group was able to generate and to verify the vast majority of reports they mapped on the Ushahidi platform. In addition, the findings from the interviews clearly show how adept DISC was at adopting new tactics in order to manage its relationship with the state. In addition, content analysis of the 2,700 reports demonstrates the high level of transparency that the project was able to achieve during the country's parliamentary elections.

The project did not significantly worsen DISC's contentious relationship with the state, although activists explained that this was due to the government being worried about possible blowback if it did crack down on the U-Shahid team. The group's connections with international allies were important, and the state could not rely on public international support for rigged elections. Despite this, the impact of Ushahidi technology on the behavior of ruling elites is less clear. As the content analysis reveals, elites did not appear to succeed in manipulating U-Shahid's independent monitoring of the elections. In terms of state capacity for repression, it is also unclear what impact the Ushahidi platform might have had. As for impact on political accessibility, the U-Shahid project had a strong positive influence, according to findings from the interviews.

The Ushahidi platform allowed DISC to circumvent state media and generate international media coverage. Meanwhile, the Egyptian regime was unable to successfully generate patriotism using social media, since it did not know how best to leverage the new media. As one activist explained "Using technology provides a comparative advantage in many ways. It makes you stand out [and] gets you lots of media coverage, free publicity. Everyone was interested in what we were doing, even political candidates and other NGOs who wanted to share their reports with us." The state was largely unable to counter the alternative frames presented by U-Shahid.

The U-Shahid project had some democratic impact on the political space and discourse in Egypt. It operated on such a small scale, however, that it is doubtful that the U-Shahid project actually succeeded in deterring election fraud in 2010. The documentation of 2,700 instances of election irregularities was impressive, but only a vastly greater number of reports could have deterred fraud.

Yet the fact that Egyptian national security was closely monitoring DISC's operations reveals that the state was concerned and treated the project as a potential political threat. The regime refrained from shutting down the project for fear of blowback. Following the fall of Mubarak in 2011, protestors stormed the offices of Egyptian national security. In the files, they found a security report on the U-Shahid project with the names and contact information (including Skype usernames) of many activists, both Egyptian and foreign, who were involved in using the Ushahidi platform.

In many ways, U-Shahid helped to reverse or at least fight back against this government-constructed panopticon, and this may have helped to pave the way for the 2011 revolution that toppled Mubarak. The Egyptian case demonstrates the value of geomapping as an important liberation technology. As John Yemma wrote:

> In Tom Stoppard's 1978 play *Night and Day*, a photojournalist in Africa notes how important it is to be able to see into dark places. 'People do awful things to each other. But it is worse in places where everybody is kept in the dark. Information is light. Information, in itself, about anything, is light.'[14]

NOTES

1. Kevin Kelly, *What Technology Wants* (New York: Viking, 2010).

2. Larry Diamond, "Liberation Technology," *Journal of Democracy* 21 (July 2010): 69–83.

3. Steven Livingston, *Africa's Evolving Infosystems: A Pathway to Security and Stability* (Washington, D.C.: National Defense University Press, 2011).

4. Clay Shirky, "The Political Power of Social Media: Technology, the Public Sphere, and Political Change," *Foreign Affairs*, January–February 2011: 28–41.

5. Philip N. Howard, "The Arab Spring's Cascading Effects," 23 February 2011, available at *www.miller-mccune.com/politics/the-cascading-effects-of-the-arab-spring-28575/*.

6. Alexey Sidorenko, "Russia: Unexpected Results of Radiation Mapping," 25 March 2011, *http://globalvoicesonline.org/2011/03/25/russia-unexpected-results-of-radiation-mapping/*.

7. Olga Werby, "Decision Scaffolding and Crisis Mapping," 6 March 2011, *www.interfaces.com/blog/2011/03/decision-scaffolding-and-crisis-mapping*.

8. A public version of this map was later made available at *LibyaCrisisMap.net*.

9. Ian Schuler, "SMS as a Tool in Election Observation (Innovations Case Narrative: National Democratic Institute)," *Innovations: Technology, Governance, Globalization* 3 (Spring 2008): 143–57.

10. Ory Okolloh, "Update Jan 3 11:00 pm," *Kenyan Pundit*, 3 January 2008 *www.kenyanpundit.com/2008/01/03/update-jan-3-445-1100-pm/*.

11. An earlier version of some of the following paragraphs was coauthored with Anahi Ayala Iacucci for the previous version of this paper, presented at the conference on "Liberation Technology in Authoritarian Regimes," Center on Democracy, Development, and the Rule of Law, Stanford University, 11–12 October 2010.

12. Unless otherwise indicated, all other quotations are from the author's interviews.

13. Philip N. Howard, *The Digital Origins of Dictatorship and Democracy: Information Technology and Political Islam* (New York: Oxford University Press, 2010), 156.

14. John Yemma, "Crowdsourcing Is Good—But Not Enough," *Christian Science Monitor*, 3 May 2011, available at *www.csmonitor.com/Commentary/editors-blog/2011/0503/Crowdsourcing-is-good-but-not-enough*.

8

EGYPT AND TUNISIA: THE ROLE OF DIGITAL MEDIA

Philip N. Howard and Muzammil M. Hussain

Philip N. Howard *is associate professor of communication at the University of Washington.* He is the author of The Digital Origins of Dictatorship and Democracy: Information Technology and Political Islam *(2010).* **Muzammil M. Hussain** *is a doctoral student in communication at the University of Washington.* This essay originally appeared in the July 2011 issue of the *Journal of Democracy.*

As has often been noted in these pages, one world region has been practically untouched by the third wave of democratization: North Africa and the Middle East. The Arab world has lacked not only democracy, but even large popular movements pressing for it. In December 2010 and the first months of 2011, however, this situation changed with stunning speed. Massive and sustained public demonstrations demanding political reform cascaded from Tunis to Cairo, Sana'a, Amman, and Manama. This inspired people in Casablanca, Damascus, Tripoli, and dozens of other cities to take to the streets to call for change.

By May, major political casualties littered the ground: Tunisia's Zine al-Abidine Ben Ali and Egypt's Hosni Mubarak, two of the region's oldest dictators, were gone; the Libyan regime of Muammar Qadhafi was battling an armed rebellion that had taken over half the country and attracted NATO help; and several monarchs had sacked their cabinets and committed to constitutional reforms. Governments around the region had sued for peace by promising their citizens hundreds of billions of dollars in new spending of various kinds. Morocco and Saudi Arabia appeared to be fending off serious domestic uprisings, but as of this writing in May 2011, the outcomes for regimes in Bahrain, Jordan, Syria, and Yemen remain far from certain.

There are many ways to tell the story of political change. But one of the most consistent narratives from civil society leaders in Arab countries has been that the Internet, mobile phones, and social media such as Facebook and Twitter made the difference this time. Using these technologies, people interested in democracy could build extensive networks, create social capital, and organize political action with a speed and on a scale never seen before. Thanks to these technologies, virtual networks materialized in the streets. Digital media became the tool that allowed social movements to reach once-unachievable goals, even as authoritarian forces moved with a dismaying speed of their own to devise both high- and low-tech countermeasures. Looking back over the last few months of the "Arab Spring," what have we learned about the role of digital media in political uprisings and democratization? What are the implications of the events that we have witnessed for our understanding of how democratization actually works today?

Tunisian Origins

On 17 December 2010, Mohamed Bouazizi set himself on fire. The young street vendor in the small Tunisian city of Sidi Bouzid had tried in vain to fight an inspector's small fine, appealing first to the police, then to town officials, and then to the regional governor. Each time he dared to press his case, security officials beat him. Bruised, humiliated, and frustrated by this cruel treatment, Bouazizi set himself alight in front of the governor's office. By the time he died in a hospital on January 4, his plight had sparked nationwide protests. The news had traveled fast, even though the state-run media had ignored the tragedy and the seething discontent in Sidi Bouzid. During the angry second half of December, it was through blogs and text messages that Tunisians experienced what the sociologist Doug McAdam calls "cognitive liberation."[1] In their shared sympathy for the dying man, networks of family and friends came to realize that they shared common grievances too. The realization hit home as people watched YouTube videos about the abusive state, read foreign news coverage of political corruption online, and shared jokes about their aging dictator over SMS. Communicating in ways that the state could not control, people also used digital media to arrive at strategies for action and a collective goal: the deposition of a despot.

For years, the most direct accusations of political corruption had come from the blogosphere. Investigative journalism was almost solely the work of average citizens using the Internet in creative ways. Most famous is the YouTube video showing Tunisia's presidential jet on runways near exclusive European shopping destinations, with on-screen graphics specifying dates and places and asking who was using the aircraft (the suggestion being that it was Ben Ali's high-living wife). Once this video appeared online, the regime cracked down on YouTube, Facebook, and other applications. But bloggers and activists pushed on, producing alternative online newscasts, creating virtual spaces for anonymous political discussions, and commiserating with fellow citizens about state persecution. With Bouazizi's death,

Ben Ali's critics moved from virtual to actual public spaces. Sham-seddine Abidi, a 29-year-old interior designer, posted regular videos and updates to Facebook. Al Jazeera used the content to carry news of the events to the world. Images of a hospitalized Bouazizi spread via networks of family and friends. An online campaign called on citizens and unions to support the uprising in Sidi Bouzid. Lawyers and students were among the first to take to the streets in an organized way.

The government tried to ban Facebook, Twitter, and video sites such as DailyMotion and YouTube. Within a few days, however, people found a workaround as SMS networks became the organizing tool of choice. Less than 20 percent of the population actively used social media, but almost everyone had access to a mobile phone. Outside the country, the hacker communities of Anonymous and Telecomix helped to cripple the government by carrying out denial-of-service attacks and by building new software to help activists get around state firewalls. The government responded by jailing a group of bloggers in early January. For the most part, however, the political uprising was leaderless in the classic sense—there was no longstanding revolutionary figurehead, traditional opposition leader, or charismatic speechmaker to radicalize the public. But there were prominent nodes in the digital networks, people whose contributions held sway and mobilized turnout. Slim Amamou, a member of the copyright-focused Pirate Party, blogged the revolution (and later briefly took a post in the national-unity government). Sami ben Gharbia, a Tunisian exile, monitored online censorship attempts and advertised workarounds. The middle-class Tunisian rapper who calls himself El Général streamed digital "soundtracks for the revolution."

By early January, urgent appeals for help and mobile-phone videos of police repression were streaming across North Africa. Ben Ali's position seemed precarious. There were major protests in Algeria, along with several self-immolations. Again, the state-run news media covered little about events in neighboring Tunisia. The Algerian government tried to block Internet access and Facebook use as traffic about public outrage next door increased. But with all the privately owned submarine cables running to Europe, Algerian authorities lacked an effective chokepoint to squeeze. When the government also became a target for Anonymous, the state's own information infrastructure suffered.

By the time Ben Ali fled Tunisia for Saudi Arabian exile on January 14, civil-disobedience campaigns against authoritarian rule were growing in Jordan, Oman, and Yemen. In other countries, such as Lebanon, Mauritania, Saudi Arabia, and Sudan, minor protests erupted on a range of issues and triggered quick concessions or had little impact. But even in these countries, opposition leaders drew inspiration from what they were tracking in Tunisia. Moreover, opposition leaders across the region were learning digital tricks for catching a ruling elite off guard. Compared to Tunisia, only Egypt had a more wired civil society, and the stories of success in Tunisia helped to inspire the largest protests that Cairo had seen in thirty years.

Egypt, Inspired

In Egypt, almost everyone has access to a mobile phone. The country also has the region's second-largest Internet-using population (only Iran's is bigger). News of Ben Ali's departure spread rapidly in Egypt, where the state-run media gave his exit grudging coverage even while continuing to move slowly on reporting the larger story of regionwide protests, including the demonstrations that were breaking out in Cairo.

Like Tunisia, Egypt has long had a large and active online public sphere frequented by banned political parties, radical fundamentalists, investigative journalists, and disaffected citizens. The state could not shut it down entirely: When the online news service of the Muslim Brotherhood (MB) was banned, for instance, servers were found in London and the organization continued to convey its views across the ether. But more than any established group, what turned anti-Mubarak vitriol into civil disobedience was a campaign to memorialize a murdered blogger.

Local Google executive Wael Ghonim started the Facebook group "We are All Khaled Said" to keep alive the memory of the 28-year-old blogger, whom police had beaten to death on 6 June 2010 for exposing their corruption. Just as digital images of Bouazizi in the hospital passed over networks of family and friends in Egypt, an image of Said's grotesquely battered face, taken by his brother as Khaled's body lay in the Alexandria city morgue, passed from one mobile-phone camera to thousands. And just as the 26-year-old Iranian woman Neda Agha-Soltan became a protest icon after her death at the hands of a regime sniper during postelection demonstrations in Tehran was caught on camera in June 2009, so did Said and his memorial Facebook page become a focus for collective dissent and commiseration. But more than being a digital tribute to someone from a group long tormented by Egyptian police, the Said Facebook page became a logistical tool, and at least temporarily, a strong source of community. Ghonim quickly emerged as Egypt's leading voice on Twitter, linking a massive Arabic-speaking social network with networks of mostly English-speaking observers and well-wishers overseas.

The first demonstrators to venture into Cairo's Tahrir (Liberation) Square on 25 January 2011 shared many hopes and aspirations with their counterparts in Tunis. They were a similar community of like-minded individuals, educated but underemployed (in a "youth-bulge" society chronically unable to create enough jobs for its legions of young people), eager for change but committed neither to religious fervor nor political ideology. They found solidarity through digital media, and then used their mobile phones to call their social networks into the streets. Protests scaled up quickly, leaving regime officials and outside observers alike surprised that such a large network of relatively liberal, peaceful, middle-class citizens would mobilize against Mubarak with such speed. Islamists, opposition-party supporters, and union members were there too, but liberal and civil society voices dominated the digital conversation. News about and speeches by Mubarak, U.S. president Barack Obama, and regional leaders were streamed live to the phones and laptops in the square.

In the last week of January, an increasingly desperate Mubarak tried to unplug his country. His attempt to cut Egyptians off from the global information infrastructure met with mixed success. Anticipating the maneuver, tech-savvy students and civil society leaders had put in place backup satellite phones and dial-up connections to Israel and Europe, and were able to maintain strong links to the rest of the world. It appears, moreover, that some of the telecommunications engineers charged with choking off Internet access were slow to move. The first large Internet service provider received the shutdown order on Friday, January 28, but took no action until Saturday. Others responded promptly, but restored normal service on Monday. For four days, the amount of bandwidth going into Egypt dropped, but it was far short of the information blackout that Mubarak had been seeking. The regime had to deal with costs and perverse effects, too. Government agencies were crippled by being knocked off the grid. And middle-class Egyptians, denied home Internet access, took to the streets in larger numbers than ever, many driven by an urge simply to find out what was going on.

A few days later, the Egyptian security services began using Facebook and Twitter to anticipate the movements of individual activists. They abducted Ghonim once his Facebook group topped three-hundred thousand members (it now has four times that many). Digital media technologies not only set off a cascade of civil disobedience across Egypt, but made for a unique means of civic organizing that was replicated around the region.

Digital media spread the details of successful social mobilization against the strongmen of Tunisia and Egypt across the region. As had happened in Tunisia and Egypt, authorities in Algeria, Bahrain, Libya, Saudi Arabia, and Syria tried to stifle digital conversation about domestic political change. These governments also targeted bloggers with arrests, beatings, and harassment. It is clear that digital media have played an important role. Images of jubilant protesters in Tunisia inspired others across the region. Facebook provided an invaluable logistical infrastructure for the initial stages of protest in each country. Text-messaging systems fed people within and outside these countries with information about where the action was, where the abuses were, and what the next step would be.

Within a few weeks, there were widely circulating PDFs of tip sheets on how to pull off a successful protest. The *Atlantic Monthly* translated and hosted an "Activist Action Plan," *boingboing.net* provided tips for protecting anonymity online, and Telecomix circulated the ways of using landlines to circumvent state blockages of broadband networks. Through Google Earth, the Shias of Bahrain—many of whom live in one-room houses with large families—could map and aggregate photographs of the ruling Sunni minority's opulent palaces. Digital media provided both an awareness of shared grievances, and transportable strategies for action.

The prominent Bahraini human-rights blogger Mahmood al-Yousif tweeted during his arrest, instantly linking up the existing networks of local democratization activists such as @OnlineBahrain with international observers through @BahrainRights. In Libya, the first assertion of a competing political authority to that of Muammar Qadhafi came online, on a website declaring an alternative government in the form of an interim national council. One of Qadhafi's senior advisors defected by tweeting his resignation and urging Qadhafi to leave Libya.

Algerians, goaded by the same sense of economic despair and dissatisfaction that drove Tunisians and Egyptians, broke out in similar demonstrations. Salima Ghezali, a leading Algerian activist, told Al Jazeera that the protests were "both very local and very global." Union-led strikes had been common in Algeria for decades, but nothing like the unrest of 2011 had been seen since 1991. Algerian protesters were not among the region's most tech-savvy, but before the country's state-run media reported on local protests or Mubarak's resignation, many residents of Algiers received the inspirational news via SMS.

Digital Contexts, Political Consequences

Ben Ali had ruled for almost twenty-five years, and Mubarak for nearly thirty. Each was tossed out of power by a network of activists whose core members were twenty-somethings with little experience in social-movement organizing or open political discourse. Seeing this, other governments scrambled to make concessions that they hoped would head off explosions. Algeria's rulers lifted an almost two-decade-old state of emergency. Oman's sultan gave its elected legislature the authority to pass laws. Sudan's war-criminal president promised not to seek reelection. All the oil-rich states committed to wealth redistribution or the extension of welfare services.

Real-world politics, of course, is about much more than what happens online. A classically trained social scientist trying to explain the upheavals would point to the youth bulge, declining economic productivity, rising wealth concentration, high unemployment, and low quality of life as common circumstances across the region. These explanatory factors are typically part of the story of social change, and it does not diminish digital media's causal contribution to note their presence. Such media were singularly powerful in spreading protest messages, driving coverage by mainstream broadcasters, connecting frustrated citizens with one another, and helping them to realize that they could take shared action regarding shared grievances. For years, discontent had been stirring, but somehow the drivers of protest never proved sufficient until mobile phones and the Web began pervading the region. It never makes sense to look for simple, solitary causes of a revolution, to say nothing of a string of revolutions, and the precise grievances have varied significantly from country to country. Yet the use of digital media to rouse and organize opposition has furnished a common thread.[2]

It is true that journalists have focused on the visible technological tactics that seemed to bring so much success, rather than looking at the root causes of social discontent. Yet this does not mean that analysts should overcorrect and exclude information technologies from the list of causes altogether. Indeed, social discontent is not something ready-made, but must gestate as people come to agree on the exact nature and goals of their discontent. In the last few years, this gestation process has gone forward via new media, particularly in Tunisia, Egypt, and Bahrain. Social discontent can assume organizational form online, and can be translated into workable strategies and goals there as well. Over the last few months, this translation process has occurred via mobile phones and social-networking applications even in countries whose governments are very good at coopting or brutally suppressing opposition, such as Saudi Arabia, Syria, and Yemen.

In the Middle East and North Africa, dissent existed long before the Internet. Yet digital media helped to turn individualized, localized, and community-specific dissent into a structured movement with a collective consciousness about both shared plights and opportunities for action. It may make more sense to think of conjoined causal combinations: the strength of existing opposition movements, the ability (or inability) of the regime to buy off opposition leaders, and the use of digital media to build opposition networks. The precise mixture of causes may have varied from country to country, but the one consistent component has been digital media.

It is premature to call these events a "wave" of democratization—their outcomes are still far too uncertain for that—yet we can say that opposition to authoritarian rule has been the consistent collective-action goal across the region. Arab social-movement leaders actively sought training and advice from the leaders of democratization movements in other countries, and rhetorical appeals for civil liberty appeared consistently from protest to protest.

As we look back over the first quarter of 2011, the story of digital media and the Arab Spring seems to have unfolded in five or perhaps six parts or phases. The first was a *preparation* phase that involved activists using digital media in creative ways to find each other, build solidarity around shared grievances, and identify collective political goals. The second was an *ignition* phase involving an incident that the state-run media ignored, but which came to wide notice online and enraged the public. Then came the third phase, a period of *street protests* made possible, in part, by online networking and coordination. As these went on, there came the *international buy-in*, during which digital-media coverage (much of it locally generated) drew in foreign governments, international organizations, global diasporas, and overseas news agencies. Matters then built toward a *climax* as regimes, maneuvering via some mixture of concession and repression, either got the protesters off the street; failed to mollify or frighten them and began to crumble before their demands; or ended up in a bloody stalemate or even civil war as we are seeing in Bahrain, Libya, Syria, and Yemen. In some cases, such as those of Tunisia and Egypt, we are seeing an additional phase of *follow-on information warfare* as the various players left standing compete to shape the future course of events by gaining control over the revolutionary narrative.

Across the region, the process of building up to political change involved "building down" the credibility of authoritarian regimes by investigating their corrupt practices. The best and perhaps the only place that critics could find for getting their message across was the Internet. Blogs, news websites, Twitter feeds, and political listservs offered spaces where women could debate on an equal footing with men, where policy alternatives could be discussed, and where regime secrets could be exposed. What set the scene for a dramatic event such as the occupation of a central square was the undramatic process of people buying cheap mobile phones or time online at cybercafés. The arrival of new digital technologies became an occasion for individuals to restructure the ways in which they produced and consumed content. When a political crisis flared, the new habits of technology use were already in place.

After 2000, new communications technologies spread rapidly across the Arab world. For many Arabs, especially in cities, reading foreign news online and communicating with friends and relatives abroad became habits. Digital media could become a near-term cause of political upheaval in 2011 precisely because they were already so popular. It may seem that digital-media use in times of political crisis is novel. But for residents of Tunis, Cairo, and other capitals, it was the sheer everyday-ness and familiarity of mobile phones that made them a proximate cause of political change. The revolution may be televised, and it is surely online.

What ignites popular protest is not merely an act of regime violence such as the police beating Mohamed Bouazizi or Khaled Said, but the diffusion of news about the outrage by networks of family, friends, and then strangers who step in when the state-run media ignore the story. When Al Jazeera failed at first to cover digital activism in Syria, civic leaders there lobbied the influential network into producing a long documentary and featuring Syrian-activist content on its website. Consequently, interest in homegrown opposition to dictator Bashar al-Assad grew rapidly both within the country and across the region.

Interestingly, the recent protest ignitions seem to have occurred without recognizable leaders. Charismatic ideologues, labor-union officials, and religious spokespeople have been noticeably absent (or at least they were at first). In Tunisia, the igniting event was Bouazizi's suicide. In Egypt, it was the Tunisian example. The rest of the region followed as scenes of demonstrators and fleeing dictators went out over Al Jazeera and social-media networks.

After ignition, the street battles of political upheaval began, albeit in a unique manner. Most of the protests in most of the countries were organized in unexpected ways that made it difficult for states to respond. The lack of individual leaders made it hard for authorities to know whom to arrest. Activists used Facebook, Twitter, and other sites to communicate plans for civic action, at times playing cat-and-mouse games with regime officials who were monitoring these very applications. In Libya, foes of the Qadhafi dictatorship took to Muslim online-dating sites in order to hide the arrangements for meetings and protest rallies. In Syria, the Assad regime had blocked Facebook and Twitter intermittently since 2007, but reopened access as protests mounted, possibly as a way of entrapping activists. When state officials began spreading misinformation over Twitter, activists used Google Maps to self-monitor and verify trusted sources. Then, too, authorities often flubbed their information-control efforts. Mubarak disabled Egypt's broadband infrastructure yet left satellite and landline links alone. Qadhafi tried to shut down his country's mobile-phone networks, but they proved too decentralized.

News coverage of events in the region regularly revealed citizens using their mobile-phone cameras to document events, and especially their own participation in them. In Tahrir Square, both the crowds and the crewmen of the tanks that were sent to watch them took pictures of one another for instant distribution to their various social networks. When army vehicles were abandoned, people clambered aboard and posed for pictures to post to their Facebook pages. Arrestees took pictures of themselves in custody. Some Egyptians speculated that the army did not act systematically against protesters because soldiers were made suddenly aware of their socially proximate connection to the square's occupants, and also because the troops knew that they were constantly on camera. In countries where the armed forces did act with aggression, including Bahrain and Syria, the resulting carnage was still documented. YouTube had to add a special waiver to its usual no-gore policy in order to allow shocking user content such as a mobile-phone video of unarmed Syrian civilians—including children—being shot by Assad's troops.

Sooner or later, regime opponents must seek some form of international support, and this, too, has become a digitally mediated process. Domestic turmoil can eventually capture international attention. Of course, the degree to which a popular uprising finds an international audience depends on strategic relations with the West, but also on the proximity of social-media networks. Most technology users in most countries do not have the sophistication to work around state firewalls or keep up anonymous and confidential communications online. But in each country a handful of tech-savvy students and civil society leaders do have these skills, and they used them well during the first months of 2011. Learning from democracy activists in other countries, these information brokers used satellite phones, direct landline connections to ISPs in Israel and Europe, and software tools for protecting user anonymity in order to supply the international media with pictures of events on the ground—even when desperate dictators attempted to shut down national ISPs.

Desperate Tactics

When conflicts between a regime and its domestic opposition come to a head, one or the other may give in, or else a stalemate (often punctuated by violent clashes) may ensue. Mounting tensions led several governments to make clumsy attempts at disconnecting citizens from the global digital "grid." Banning access to social-media websites, powering down cell towers, or disconnecting Internet switching points in major cities were among the desperate tactics to which authoritarian regimes resorted as they strove to maintain control. Even short disruptions of connectivity were costly. Egypt lost at least US$90 million to Mubarak's only partly successful efforts to cut off digital communications. Perhaps even more damaging in the long run, this episode harmed the country's reputation among technology firms as a safe place to invest. In Tunisia, the situation was reversed: It was not the government but rather activist hackers—or "hacktivists," as many call themselves—who did the most economic damage by shutting down the national stock exchange.

When regimes struck back, their counterblows had digital components. Bahrain, Morocco, and Syria saw cyberstruggles to dominate Twitter traffic. Every country that experienced turmoil witnessed delays or disruptions in mobile-phone and Internet service, but it is hard to say whether this was due to regime-driven shutdowns or overwhelming traffic volumes at moments of maximal uncertainty. Quite likely it was both. The zenith of crisis tended to mark the nadir of connectivity as regimes cracked down on large telecommunications providers while skyrocketing traffic was rerouted to a few small available digital switches.

The information wars that followed the protests of the Arab Spring began with the efforts of regimes to hide their tracks. In Egypt, the State Security Investigations Service—Mubarak's political police—did all it could to destroy its archives, though some records leaked online. The websites of activists, meanwhile, became portals for criticisms of the interim government and its leaders.

The victors in a popular uprising generate ever more digital content, while the supporters of failed dictators produce less and less. Deposed dictators find only a small audience online, while the entrepreneurial activists who served as important nodes in the social-movement network find themselves with newfound leadership roles. By the time the protests are over, a few of the "digerati" such as Wael Ghonim find that they have become newly prominent public figures. And the public expectation of being able to use information technology to access political figures remains. When U.S. secretary of state Hillary Clinton was booked for a Web chat with a popular Egyptian website, around 6,500 questions were submitted in just two days.

Traditional media sources also played an important role in the Arab Spring. Satellite television forged a strong sense of transnational identity across the region, and everyone recognized the importance of coverage in this medium: Both Mubarak and his information minister called television anchors personally to berate them for unflattering stories. Of the existing news organizations, Al Jazeera certainly enjoyed the highest profile and the most influence regionally. The network's Dima Khatib, a native of Syria, was the most prominent commentator on Tunisia when that country erupted, and she served as a key information broker for the revolution through her postings on Twitter. Al Jazeera had an exceptionally innovative new-media team that converted its traditional news product for use on social-media sites and made good use of the existing social networks of its online users. But a key aspect of its success was its use of digital media to collect information and images from countries in which its journalists had been harassed or banned. These digital networks gave Al Jazeera's journalists access to more sources, and gave a second life to their news products. Indeed, the use of social media itself has become a news peg, with analysts eager to play with the meme of technology-induced political change.

Regime responses varied in sophistication, but often seemed several technological paces behind the behavior of civil society. In February, during one of his televised speeches, Qadhafi interrupted his train of thought when an aide drew his attention to real-time coverage of his rant. Qadhafi had simply never encountered such instant feedback from a source that could not easily be silenced or punished. In Bahrain, the successful suppression of protest by the country's Sunni-dominated monarchy gave it an opportunity to plug the security holes in its telecommunications network. Though never as severely challenged by demonstrators, Saudi Arabia's rulers have reorganized the server infra-structure of the Kingdom so that all Internet traffic there flows through exchange points that are physically located in Riyadh.

It is a mistake to build a theory of democratization around a par-ticular kind of software, a single website, or a piece of hardware, or to label these social upheavals "Twitter Revolutions" or "WikiLeaks Revolutions."[3] Nor does it make sense to argue that digital media can *cause* either dictators or their opponents to achieve or fall short of their goals. Technological tools and the people who use them must together make or break a political uprising.

The Digital Scaffolding for Civil Society

Digital media changed the tactics of democratization movements, and new information and communication technologies played a major role in the Arab Spring. We do not know, at the time of this writing in May 2011, where events in the various countries will lead, and whether or not change will come to the remaining, more recalcitrant authoritarian governments. But the consistent narrative arc of the uprisings involves digital media. The countries that experienced the most dramatic pro-tests were among the region's most thoroughly wired, and their societies boasted large numbers of people with the technical knowledge to use these new media to strong effect.

In times of political crisis, technology firms may "lean forward" with new tools or applications introduced to serve an eager public (and in doing so, capture market share). In late January, for example, Google sped up its launch of its speak-to-tweet service, an application designed to translate voicemails into tweets as a means of bypassing Mubarak's Twitter blockade. Several tech firms built dedicated portals to allow in-country users to share content. But as Evgeny Morozov has pointed out, information technologies—and the businesses that design them—do not always end up supporting democratization move-ments.[4] Opposition leaders in countries where political parties are ille-gal sometimes use pseudonyms to avoid government harassment. But doing so on Facebook is a violation of the company's user agreement, and so the company actually shut down one of the protest-group pages in December. Supporters eventually persuaded Facebook to reinstate the page, but the incident showed how businesses such as Facebook, YouTube, and Twitter may not fully appreciate the way in which their users treat these tools as public-information infrastructure, and not just as cool new applications in the service of personal amusement. Whereas Google has signed the Global Network Initiative—a compact for preventing web censorship by authoritarian governments—Face-book has refused to do so. It might be technically possible to require Facebook users in Western countries to use real identities while allowing people living under dictatorships to enjoy anonymity, but no such feature currently exists.

Absent digital media, would the Arab Spring still have occurred? It is hard to say. The Arab world has long had democratic activists, but never before had any toppled a dictator. Radio and television have long reached most Arabs, but only 10 to 20 percent of those living in a typical Middle Eastern or North African country can easily gain access to the Internet. Yet this minority is a strategic one, typically comprising an elite made up of educated professionals, young entrepreneurs, urban dwellers, and government workers. These are the people who formed the networks that initiated, coordinated, and sustained successful campaigns of civil disobedience against authoritarian rule. Looking at the other side of the coin, the countries with the lowest levels of technology proliferation have also tended to have the weakest democratization movements. As fascinating as it can be to think of counterfactual scenarios, it would be a mistake to see these as belonging on an equal footing with actual events in concrete cases concerning which we have ample empirical evidence. Counterfactuals and thought experiments can be fun, but in the search for patterns that is the social scientist's task, prominence should always be given to real cases and the real evidence they yield.

As we have noted, it is premature to assert that we are witnessing a wave of democratization. Several states are still in crisis. In countries where authoritarian governments have collapsed or made major con-cessions, it is hard to know whether stable democracies will emerge. Democratization waves are measured in years, not months. In 1998, Indonesia's Suharto fell when students using mobile phones success-fully mobilized and caught his regime off guard, but it took a decade of difficult political conversations for democratic practices to become entrenched. The Arab Spring had a unique narrative arc, involved a par-ticular community of nations, and caught most autocrats and analysts alike by surprise. Digital media are important precisely because they had a role in popular mobilizations against authoritarian rule that were unlike anything seen before in the region.

It is also noteworthy that a remarkable amount of political change has occurred in a surprisingly nonlethal manner. In Algeria, Egypt, Jordan, Morocco, and Tunisia, civil society leaders found that the security ser-vices showed a remarkable reluctance to move aggressively against pro-testers (and the Tunisian and Egyptian militaries did not want to move against them at all). Could this hesitancy have had anything to do with the large numbers of mobile-phone cameras that demonstrators were carrying? Sadly, a distaste for the use of deadly violence by regime forces has not been evident in Bahrain, Libya, Saudi Arabia, Syria, and Yemen. Yet even in those cases, it can at least be said that solid docu-mentation of regime abuses or even atrocities has reached the interna-tional community, in no small part thanks to mobile phones.

Scholars of social movements, collective action, and revolution must admit that several aspects of the Arab Spring challenge our theories about how such protests work. These movements had an unusually wide or "distributed" leadership. The first days of protest in each country were organized by a core group of literate, middle-class young people who had no particular affinities with any existing political parties or any ideologies stressing class struggle, religious fundamentalism, or pan-Arab nationalism. Broadcast and print media—long associated with the mobilization phase of democratization waves—took a decided backseat to communication via social networks. This communication, moreover, itself had a strongly distributed or lateral character and did not consist of one or a few relatively simple ideological messages beamed by an elite at a less-educated mass public, but had more the character of a many-sided conversation among more or less equal individuals.

Seeing what has unfolded so far in the Middle East and North Africa, we can say more than simply that the Internet has changed the way in which political actors communicate with one another. Since the beginning of 2011, social protests in the Arab world have cascaded from country to country, largely because digital media have allowed communities to unite around shared grievances and nurture transportable strategies for mobilizing against dictators. In each country, people have used digital media to build a political response to a local experience of unjust rule. They were not inspired by Facebook; they were inspired by the real tragedies *documented* on Facebook. Social media have become the scaffolding upon which civil society can build, and new information technologies give activists things that they did not have before: information networks not easily controlled by the state and coordination tools that are already embedded in trusted networks of family and friends.

NOTES

1. Doug McAdam, *Political Process and the Development of Black Insurgency, 1930–1970* (Chicago: University of Chicago Press, 1982).

2. Philip N. Howard, *The Digital Origins of Dictatorship and Democracy: Information Technology and Political Islam* (New York: Oxford University Press, 2010).

3. Elizabeth Dickinson, "The First WikiLeaks Revolution?" Foreign Policy Online, available at *http://wikileaks.foreignpolicy.com/posts/2011/01/13/wikileaks_and_the_tunisia_protests*. See also Andrew Sullivan, "Tunisia's Wikileaks Revolution," Atlantic Online, available at *www.theatlantic.com/daily-dish/archive/2011/01/tunisias-wikileaks-revolution/177242*.

4. Evgeny Morozov, *The Net Delusion: The Dark Side of Internet Freedom* (New York: PublicAffairs, 2011).

9

CIRCUMVENTING INTERNET CENSORSHIP IN THE ARAB WORLD

Walid Al-Saqaf

Walid Al-Saqaf *is a Yemeni journalist, software developer, activist and media scholar, currently directing the Masters in Global Journalism program at Örebro University in Sweden. Al-Saqaf is the founder of Yemen Portal, a news aggregator, and Alkasir, a software program aimed at mapping and circumventing censorship worldwide.*

In mid-January 2011, successive and massive protests throughout Tunisia forced out President Zine al-Abidine Ben Ali, one of the Arab world's most repressive autocrats. Scholars labeled this dramatic event the Jasmine Revolution. Others emphasized social networking's influence on Ben Ali's exile, and titled the event the Facebook Revolution.[1] This title is also apropos because there are approximately two million Tunisian Facebook users, roughly 19 percent of the country's population.[2] This made local Facebook users a loud, clear, and formidable voice of the opposition. However, the Tunisian phenomenon would be best described as a "network revolution," since it was contingent on youth and activists utilizing a variety of social-networking tools, not just Facebook, through a combination of online, cell-phone, and in-the-flesh social gatherings.

Remarkably, this network revolution occurred in an Arab country that aggressively censored the Internet under Ben Ali's direction for many years. In his refuge in Jeddah, Saudi Arabia, the ousted president is likely asking himself, "What went wrong?" It would be understandable for Ben Ali to pose such a question, especially since advisors led him to believe that Tunisia's Internet was under his control, preventing opposition forces from organizing and rallying against him. However, the Tunisian regime underestimated the resilience and innovation of Arab bloggers and cyberactivists when it came to using technology to overcome the effects of Internet censorship. Activists such as Sami Ben Gharbia, an exiled Tunisian journalist, always believed that the Internet could facilitate a transition from dictatorship to democracy despite government censorship. In a tweet following the revolution, Ben Gharbia said, "we won't be silenced."[3] In short, Tunisia was a battleground between those who censored and those who circumvented. In the end, the latter had the upper hand.

Luckily, the story of circumventing Internet censorship does not end with Tunisia. Within a few weeks of the Tunisian uprisings, demonstrators flooded Egyptian streets on 25 January 2011, the so-called "Day of Wrath." Egyptian cyberactivists used Facebook and Twitter to coordinate demonstrations, triggering a series of massive street protests in which more than a million Egyptians participated.[4] Many of the protestors chanted for the end of President Hosni Mubarak's three-decade-long rule. Authorities quickly resorted to censorship once demonstrations in Egypt began. They jammed mobile-network traffic and blocked access to Twitter and Facebook.

Soon after this censorship began, several Egyptians contacted me asking for methods to circumvent it. With my circumvention software, they succeeded in overcoming Mubarak's blockades. A Twitter search for "#jan25 and proxy" on January 26 showed that many Internet users had already begun sharing information about circumvention methods. The January 25 rally demonstrated that Egyptians had to use circumvention technologies, as the Tunisians had, to help keep users connected and networked through the various online social-networking platforms. Perhaps realizing that Internet censorship alone was not effective enough to stop the flow of information, the Egyptian regime took the extreme approach of completely shutting down the Internet for five days, from January 27 to February 2.

The Tunisian and Egyptian cases show how circumvention methods and tools for bypassing Internet-censorship can reduce its impact by maintaining an unimpeded flow of information and interaction among users. This technology has great importance and relevance to democracy in the Arab world. Utilizing an interdisciplinary approach that draws on both information technology and the social sciences, this chapter analyzes Internet censorship patterns as well as the effects and use of circumvention tools in the Arab world.

How It All Started

Although Internet censorship comes in various forms,[5] I focus here on a common form of censorship that limits Internet users' access to certain online content: filtering. Internet service providers (ISPs) usually practice website filtering, but it may also be done on a local router serving as a gatekeeper that monitors page requests and checks them against a blacklist of addresses and keywords.

Before delving into the core findings of this study pertaining to filtered content and circumvention in the Arab world, it is useful to provide some background on Internet censorship in Arab countries and the origins of this research.

First, it is important to remember that Tunisia and Egypt were not the only countries blocking websites from their citizens. Most repressive Arab regimes restrict access to websites and services, and of course Internet censorship is also practiced widely by a variety of other countries, including China, Thailand, and Iran.[6]

Second, Internet censorship diminishes access to information and thus the spread of democratic ideals. Ironically, for many (including myself), exposure to Internet regulation opened individuals' eyes to the previously mysterious field of Internet censorship. Observing censorship motivated me to study this phenomenon and develop mechanisms to circumvent it. Furthermore, evading persistent censorship hones the skills of cyberactivists and software developers in their battle against repressive governments.

Third, a disclaimer: I am not neutral on the question of Internet censorship. I oppose it. I have been a victim of censorship myself and know how frustrating it is to find thousands of your online readers suddenly cut off from your website. Having been confronted with Internet censorship in the Arab world, the issue resonates with me personally. Hence, I feel that it is appropriate for me to provide a brief background about myself, and to explain why I feel the need to pursue the path of studying and circumventing Internet censorship.

In 2007, I launched a unique online-content aggregator and search engine for my native country of Yemen. I named it "Yemen Portal" *(yemenportal.net)* because I wanted it to be a window to everything that had to do with Yemen, whether it was news, opinion articles, videos, blog posts, or other discussions. I wanted to ensure that individuals would have an equal opportunity to access all sources, whether pro-government, pro-opposition, or independent. Google News inspired me to build Yemen Portal to be a place where readers had a panoramic view of information through a spectrum of news sources and political parties. Yemen Portal attempted to combat the often-limited perspective shown on monopolized TV, radio, and print media. I deliberately launched the website on the seventeenth anniversary of Yemen's unification, 22 May 2007.[7]

From June 2007 to January 2008, there were more than 200,000 visits (including 68,000 unique visits) to Yemen Portal from the world over. Interestingly, visitors to the site clearly preferred nongovernmental or oppositional and independent content, even though the average number of published articles per day from government websites was higher.[8] In fact, readers' first preference was oppositional content, while independent media was their second choice, and governmental content a distant third. I openly shared my results with website administrators and online journalists involved in opposition, government, and independent websites. Soon after, on 19 January 2008, the government of Yemen censored my website by means of a firewall. Having my website filtered was a pivotal moment in my career and motivated me to study Internet censorship and find ways to circumvent it not only in Yemen, but also in the wider Arab world. When my university supervisor approved my doctoral dissertation to study Internet censorship in the Arab world, I felt that it had become both my personal and professional mission to study this problem with an eye toward finding solutions to it.

In order to identify which websites Arab regimes blocked from users, I developed a software solution in May 2009 named the "Alkasir Solution for Internet Censorship Mapping and Circumvention." Since the website was a homegrown project intended to help local Arab Internet users, I gave it the name "Alkasir" which means "the circumventor" in Arabic. I also made the website multilingual with an Arabic and English interface.

Alkasir, like other circumvention technologies, works quite simply. It loads a list of website addresses that are known to be censored within a specific country into its central database. This list varies based on the country in which the users are located and the ISP they use. It also allows users to report a website or address that they suspect may be blocked in their country. The system verifies whether the website is blocked by comparing a locally retrieved version with one retrieved through a remote server located where there is no censorship. Once this verification is complete, Alkasir moderators check to see if the website meets the software's submission policy, which states that all websites can be submitted except those with nudity, pirated material, or viruses and malware. Once approved, the previously blocked websites become accessible not only to the specific user who submitted the link, but to all Alkasir users in that country where the link was censored.

I have received some criticism for not allowing content such as pornography to be accessible. In a radio interview published on *Voice of America,* a commentator questioned whether I was also guilty of censorship because I was not allowing certain content.[9] The interviewer compared Alkasir to other well-established circumvention solutions available worldwide such as Tor, Ultrasurf, and Hotspot Shield, which allow pornography sites. I would offer two rebuttals to this critique: one technical, the other philosophical. First, it is important to understand that I tailored my circumvention tool primarily for Arab users who are unable to access blocked social networks, blogs, news, and other online information. The purpose of the tool is to help users connect with others through Facebook, receive and transmit information through blogs, and read different perspectives of news from various parties, including dissident sources. By directing Alkasir's server and bandwidth resources to information content rather than massive and bandwidth-hungry pornographic videos, individuals use the resource more efficiently and effectively for the intended purpose. It is also important to highlight that Alkasir can be used to access websites that offer additional circumvention tools that can be used to access any content. Importantly, Alkasir is simply not a filtering system, because when an individual installs Alkasir it does not prevent access to previously accessible websites. Additionally, due to technical constraints, including costly hardware, it would have been impossible for Alkasir to circumvent all blocked content.

Second, the political and philosophical aim of Alkasir stems from the belief that the Arab world lacks real democracy, which Arab peoples deserve just as much as people anywhere else in the world.[10] For this reason it is crucial to prioritize news and information over "adult content." Also, in order to appeal to Arab users (who are generally socially conservative) and convince them of the positive contribution of Alkasir, it had to be clearly stated that it does not facilitate viewing pornographic websites. Furthermore, it is advantageous not to allow pornographic sites because an Arab-based circumvention solution allowing this content would provide ammunition to progovernment and religious figures in the region.

It is worth noting that the project was made possible by utilizing open-source code available for free online. The software is still exclusive to Windows-based PC platforms, but there are plans to expand to other operating systems.

Alkasir basically works by having "tunnels" access blocked websites' traffic through proxy servers that use a split-tunneling technique inspired by Virtual Point Network (VPN) services. A VPN is basically a relay point or proxy through which remote users can access the Internet. A designated tunneling proxy, which different Internet Protocols (IPs) can access based on certain variables, allows Alkasir to preserve bandwidth on the proxy server while actively circumventing censored content.[11]

Use and Demographics

Between 17 September 2010[12] and 26 January 2011, a total of 7,071 users from 72 countries and territories accessed Alkasir.[13] Visitors from the Arab world constituted the majority at 68 percent with 4,821. As we see from Figure 1, most Arab Alkasir users during this period were from Syria, Saudi Arabia, and Yemen. Users from the three countries totaled 3,812 people (79 percent of all Arab users).

It is important, however, to acknowledge that although there were many Arab users during the five-month period analyzed, not all of them accessed the service simultaneously. Based on 26 January 2011 statistics, on average around 800 users used Alkasir daily. Furthermore, the distribution of access during the day shows that peak usage usually

FIGURE 1—ALKASIR USERS IN THE ARAB WORLD

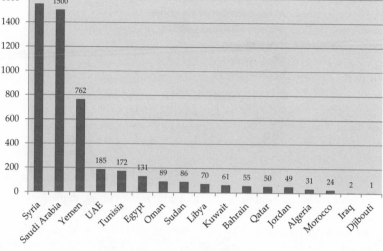

Note: Covers the period from May 2010 to January 2011.

occurs between 1500 and 2200 hours Greenwich Mean Time, with an average of 330 users connected to the server simultaneously.

It is not necessary to be a member of alkasir.com to use Alkasir, nor is it necessary to download and install Alkasir in order to become a member. Nonetheless, 5,463 members had joined alkasir.com by 20 January 2011. It should be acknowledged that in order to protect the privacy of Alkasir users, the server does not archive Alkasir users' IP addresses. Alkasir.com also embeds anonymity within the system since users do not need to provide their real name or country.

Arab Regimes Embracing Internet Censorship

Alkasir's data confirms that many Arab governments have attempted to block information on the Internet. Based on data collected from Alkasir clients in the Arab world from 10 January 2010 to 28 January 2011, many Arab state-run ISPs were found to have filtered Uniform Resource Locators (URLs) and websites containing various content, especially social-networking and political websites. The data proving this censorship is stored in a central database in a U.S.-based server.

By 31 January 2011, the Alkasir vetting system placed 501 blocked URLs into 21 categories, based on the content of the websites as determined by me and Alkasir volunteers. (However, there were dozens of additional websites that were not yet categorized and were not used in this study. This could affect the analysis here as to which genre of website Arab governments choose to block most consistently.)

FIGURE 2—DISTRIBUTION OF BLOCKED URLS IN THE ARAB WORLD

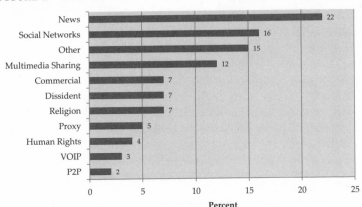

News websites topped the list of blocked content, representing 22 percent of all filtered URLs (Figure 2). Social-networking websites accounted for 16 percent of all blocked websites. The third most blocked category of website was multimedia sharing, which made up about 12 percent of all blocked URLs. Every other specific category represented a percentage of blocked websites of less than 10 percent.

The distribution of blocked content is reminiscent of censorship practices targeting traditional media by an authoritarian regime, such as the one in Yemen. Alkasir's website categorization was based on the service or content offered via the different censored websites. Dissident views, criticisms of the government, reports of human-rights violations, and similar content historically censored by autocratic governments in traditional media certainly figure among the most frequently blocked online content. This clearly demonstrates that censorship patterns transcend the medium carrying the information. Governments have heavily invested in firewall software to suppress news and opinion content, social networks, and multimedia sharing. The fact that those three groups constitute over 49 percent of all blocked URLs indicates how keen Arab regimes are to keep their information-suppression techniques up to date.

From 10 January 2010 to 31 January 2011, Alkasir recorded 13,159 cases from Arab countries where the blocked URL met its approval policy by not including pornography, viruses, or pirated content. Each of those entries represented an attempt of an Alkasir client to access a URL. This reporting process was carried out either manually by having the user type in a URL in a designated input box, or automatically through a built-in script that regularly cross-validates the status of URLs reported by Alkasir users in the same country.

It is important not to confuse the number of verified cases of URL blocking with the number of blocked URLs. The number of veri-

TABLE—URL BLOCKING IN ARAB COUNTRIES

Country	ISPs	Blocked URLs
Syria	52	177
Saudi Arabia	62	175
Yemen	12	84
U.A.E.	18	45
Bahrain	20	44
Libya	3	11
Sudan	7	7
Egypt	14	6
Jordan	16	4
Qatar	9	3
Kuwait	8	3
Total	**221**	**559**

fied URL blocking cases refers to the times a user discovered that an ISP in a country blocked a URL. Based on this understanding, it is possible for one URL to have dozens of entries testifying to blocking. Furthermore, in the same country, there could be many individuals reporting the same URL as blocked, creating numerous verified but redundant cases for the same URL. For example, if a country has 200 ISPs reporting an identical URL, this would indicate 200 entries in the database but only one blocked URL.

The study found 413 blocked URLs of political, multimedia-sharing, and social-networking content, across eleven Arab countries. Our data documents the Arab countries that practice blocking, the number of blocked URLs, and the number of ISPs that were found to practice such blocking as of 31 January 2011 (see the Table). This table does not mean that each of the ISPs blocked all the censored URLs, however. Rather, the table reports the number of ISPs that blocked at least one URL, and of course some ISPs blocked a great many more. There could be cases where only one ISP blocks most of the URLs, while on other occasions many ISPs may block most of the URLs. However, it is important to note that a number of the governments conducting censorship represent highly authoritarian regimes (Saudi Arabia, Yemen, and Syria, for instance) that tend to monopolize the Internet-service sector or to tightly control Internet access through restrictive laws and regulations.[14]

Syria, which blocked 177 URLs, was the foremost Internet filterer and was followed closely by Saudi Arabia with 175 blocked URLs (see Table 1). Yemen, the United Arab Emirates, Bahrain, and Libya also practiced censorship to a lesser extent. Blocking of a few websites was detected in Sudan, Egypt, Jordan, Qatar, and Kuwait. It is important to note that Egypt only began blocking websites on 25 January 2011, when it filtered some social-networking websites. These activities continued until 28 January 2011, when the Egyptian government cut off citizens' access to the Internet completely. There was no reported blocking by other Arab countries.

News and social-networking sites were blocked by the highest number of countries. Nine Arab countries blocked news websites because the information these websites contained could not be controlled through license revocation, pre-publishing censorship, or other conventional censorship techniques. In all Arab countries represented in the study, there are press laws whose effect is to keep media from publishing freely, sometimes by means of politically motivated red tape.[15] Authoritarian Arab regimes justified Internet censorship as needed to protect national security, religious sensitivities, and social norms. These are rationalizations for limiting freedom of speech in print and broadcast media.

Traditional press laws are largely irrelevant to Internet news because online content is immediate and decentralized. Although it remains to be empirically verified, it appears that factors behind blocking news websites are quite similar to those behind blocking other politically oriented dissident, human-rights, minority-related, and reformist websites.

Seven Arab countries blocked dating websites, which could be deemed offensive to religious or cultural traditions. Perhaps the autocrats blocked the websites to maintain conservative social values.

Six countries, especially those with strict religious stances, blocked websites devoted to minority faiths. Governments often describe these religious websites as insulting to Islam.

Some countries also blocked multimedia-sharing websites such as YouTube. Since multimedia-sharing websites and blogs are similar to social-networking sites—all strengthen users' networks by enabling them to form groups and frame plans for "real-world" actions—there is a similar rationale behind blocking them. Such websites are the bedrock of network revolutions like those witnessed in Tunisia and Egypt, making it natural for oppressive Arab regimes to take aim at social-networking. Furthermore, many social-networking sites are increasingly including multimedia-storage mechanisms. For example, Facebook allows users to upload videos and pictures. Similarly, YouTube has adopted features associated with social networking, including a flexible profile, a comment section, and a messaging service. It seems apparent that the two website categories are slowly converging.

Six Arab countries blocked proxy and circumvention websites due to the websites' capacity to allow users to evade censorship and negate the effects of filtering. Five countries blocked blogging platforms. As with social-networking websites, regimes filter blogs because of their inherent ability to allow users to generate and publish unrestricted content.

While this categorization of blocked content is useful and illustrative on its own, it remains difficult to draw conclusions, solely from the statistics, regarding why different countries are blocking specific categories and without doing extensive research into each particular country's rationale. Some countries block numerous forms of online political content, though they have different priorities. For example, Libya appeared interested only in blocking news and opposition websites (e.g., akhbar-libyaonline.com, almanaralink.com)[16] that specifically targeted the Libyan regime of the late Muammar Qadhafi. On the other hand, Bahrain and Yemen aggressively censored all sorts of websites that had political content, including blogs, human-rights reports, and news sites.[17]

Do Arabs Support Web Censorship?

Just as Alkasir users actively reported blocked URLs from the Arab world, they also contributed to this chapter by answering an online survey regarding Internet censorship and circumvention.

Alkasir developed an online survey with 71 questions. The survey was published in both English and Arabic on Alkasir's website on 8 June 2010. Survey questions were formulated to make use of results from previous research. For example, questions on censorship were split into technical and nontechnical censorship.[18] Furthermore, researchers used information from the study by the Global Internet Freedom Consortium[19] when designing the questions. After researchers received answers to the questions, they conducted quantitative content analysis.

Although Alkasir sent invitations to fill out the survey[20] through channels ranging from email to subscription lists, the majority of respondents were Alkasir users. They included users who are not only interested in viewing online content, but also supposedly eager to pursue methods to circumvent censorship. Upon installing Alkasir for the first time, the user is prompted to fill in the survey in order to activate the program. For this study, analysis considered only the completed survey results and questions relevant to research questions.

By 31 January 2011, the survey had a completion rate of about 74 percent (and in turn 70 percent of the respondents were from Arab countries).[21] Not surprisingly, there is a positive correlation between the number of respondents in a country and the number of URLs blocked in each country. The only exception to this finding is Yemen, which had a higher number of respondents due to the extra publicity that Alkasir received there through the Yemen Portal website. Tunisia had no websites reported as blocked since most respondents completed the survey after the fall of the Ben Ali regime in early 2011, which led to the end of Internet censorship there.

Analysis of the surveys confirmed the hypothesis that Arab users were strongly opposed to blocking certain content. This was understandable because the survey respondents were those using circumvention software. Of those surveyed, 69 percent of the respondents stated that it was always or mostly inappropriate to block political content, while 67 percent stated that blocking circumvention websites was always or mostly inappropriate. Yet respondents did not mind the blocking of other forms of content such as pornography, security-related, and gambling websites. When compared to results from surveys from non-Arab countries, Arabs were much more opposed to the blocking of political content, while non-Arab respondents were more opposed to the filtering of copyrighted and hacking content.

When the survey asked Arab users whether they believe censorship of certain content is appropriate, a pattern emerged whereby Arab users voiced support for censorship of pornographic and adult content while they strongly opposed censorship of news content.

About one of every seven of our Arab survey respondents (14 percent) were bloggers or administrators of websites. Nearly a third of the survey respondents had their website or blog blocked at least once, while another 8 percent believed that their blog or website would be blocked eventually. Arab respondents who had suffered from governments blocking their blogs or websites, as well as those who had indicated that their country practiced filtering, agreed that Internet censorship had negatively affected their online experiences. Slightly more than two-thirds (68 percent) said that filtering inhibited democratic expression, while 64 percent said that it inhibited social networking and also accessing information from diverse sources.

The diminished online experience that results from government restrictions on web content may explain why 79 percent of those respondents actively use circumvention tools. Survey results indicate that respondents use a variety of circumvention solutions ranging from web-based proxies to VPN services. The most common method of circumvention used by this demographic was Hotspot Shield, which was used by 35 percent, followed by Ultrasurf with 24 percent, and a variety of web-based proxies each with less than 18 percent. All these solutions are available free of charge, but Hotspot Shield is a VPN client allowing the circumvention of a broader number of Internet-censorship techniques, which may explain why it topped the list.

In the Wake of the Arab Spring

Alkasir usage surged in 2011 as compared with 2010. The reason may be largely attributed to the Arab Spring, which resulted in more awareness of the Internet as an effective means of delivering messages calling for political change and organizing dissident movements.

Based on conversations with some Alkasir users in the Arab countries that are undergoing revolutions, online censorship was viewed as the first and biggest hurdle to overcome in order to publish politically challenging text and multimedia content and to report on rallies, demonstrations, and human-rights violations that the mainstream media may not be able to cover directly.

In Syria, for example, Alkasir witnessed a tremendous rise in the number of installations, from a few thousand in 2011 to more than 28,500 in January 2012. The software's popularity in Syria as a circumvention tool grew steadily after it was discovered by some Egyptian activists trying to circumvent censorship of Twitter.com before the Internet was shut down at the beginning of the Egyptian revolution.[22]

During 2011, a number of Alkasir users shared with us by e-mail their reliance on Alkasir for circumventing censorship. Some Syrian users said that they had tried various circumvention solutions but found Alkasir to be among the more reliable and less prone to bottlenecks and disruptions. This may be partly due to its efficient split-tunneling approach, which sets it apart from other similar solutions.

As of January 2012, Alkasir had been run approximately three-million times by more than 67,500 users in 116 countries. Almost half these users (slightly more than 30,000) were from Arab countries, and of these, more than two-thirds were from Syria (with over 21,074 installations), followed by Yemen and Saudi Arabia with about 3,000 each.

By January 2012, there were slightly more than a hundred-thousand reports of blocked URLs, which consequently allowed Alkasir users to circumvent the censorship of 3,042 unique URLs in total (about half in various Arab countries). Not surprisingly, Syria came out on top in terms of the number of blocked URLs with 734, followed by Yemen with 318 and Saudi Arabia with 240.

By January 2012, 651 of the blocked URLs had been categorized.[23] Among those URLs, more than a third, or 232, belonged to the news category (but among the blocked URLs in Syria, the proportion of news sites was even higher, topping 40 percent). The second-largest category was social networking with 148 URLs (23 percent) followed by multimedia with 98 (15 percent). The other categories with a significant number of URLs were religion and politics (7 percent each), blog platforms (5 percent), human rights (4 percent), and minority faiths (2 percent).

A good indicator of what users were interested in is the number of times that a particular website was reported as blocked. It was found that users' interests varied widely depending on the country in which they were based. If we lump all users together, arabtimes.com, a satirical news website exposing scandals involving Arab officials and celebrities, was the URL most frequently reported as blocked. It was followed closely by Facebook, the world's leading social-networking website. Other websites frequently reported ranged from censorship-circumvention services such as hotspotshield.com to human-rights websites such as anhri.com.

Upon closer examination of usage patterns in Syria, it was found that the most commonly reported blocked URL was efrin.net, a Kurdish website promoting the rights of the Kurdish minority in Syria and supporting the Syrian popular revolution calling for the ouster of President

Bashar al-Assad. The second most commonly reported blocked URL was Facebook. From communicating with Alkasir users in Syria, it was clear that accessing Facebook was of great importance to them. The second position was also shared by 2shared.com, a multimedia content-sharing website often used to upload video and images. Other websites that were reported often were Kurdish websites and regional news websites such as the Lebanese daily annahar.com. Multimedia websites such as YouTube and social-networking sites such as couchsurfing.com were also frequently reported as having been blocked.

A conclusion that could be drawn from the usage of Alkasir in Syria is the apparent focus of users there on websites exposing injustice to minorities, such as Kurds. But more generally, Syrian Internet users wanted to express themselves freely and connect with one another, which is why they sought unimpeded access to social-networking and multimedia-sharing websites.

Since the Arab Spring began, some Alkasir users have reported that ISPs in Saudi Arabia and Syria have fine-tuned their filtering software to render some circumvention tools ineffective. To confront such challenges, new updated versions of Alkasir have used obfuscation in an attempt to prevent ISPs from identifying traffic patterns connecting users.

Internet censorship—affecting news, opinion, and social-networking sites—is pervasive in many Arab countries. But Arab bloggers and Internet users are not passively accepting it. The majority of Arab bloggers using Alkasir also used additional circumvention technologies. It is possible to infer from this that Internet censorship is not effective in preventing them from reaching or publishing online content.

Although circumvention may have helped empower users to escape autocratic Arab regimes' complete control of information, it certainly is not a magic bullet. It has several inherent shortcomings such as lax security and slower connection speeds. Furthermore, circumvention is not helpful when Internet access is completely shut down, as it was for a brief time in Egypt. Although many remain hopeful that liberation technology will change the region, more research needs to be conducted to understand the long-term effects of Internet filtering and circumvention tools. But this much is clear: If the current trend of increasing Internet access and online activism continues, the familiar cat-and-mouse game between ISPs in authoritarian Arab states on the one hand and censorship-circumvention tool developers and users on the other will probably continue for years to come.

NOTES

1. Mike Giglio, "Tunisia Protests: The *Facebook* Revolution," *Daily Beast*, 15 January 2011; available at *www.thedailybeast.com/blogs-and-stories/2011-01-15/tunisia-protests -the-Facebook-revolution.*

2. Jaco Maritz, "Top 10 African Countries on *Facebook*," *How We Made It In Africa*, 19 January 2011; available at *www.howwemadeitinafrica.com/top-10-african-countries-on-Facebook/6980*.

3. Sami ben Gharbia, Twitter post, 18 January 2011, 4:15 p.m.; available at *Twitter.com/#!/ifikra/statuses/27474214818676736*.

4. Kiana Ashtiani, "Egyptian Uprising: A Timeline," *prospectjournal.ucsd.edu*, 28 February 2011; available at *prospectjournal.ucsd.edu/index.php/2011/02/egyptian-uprising-a-timeline*.

5. Walid Al-Saqaf, "Internet Censorship Challenged: How Circumvention Technologies Can Effectively Outwit Governments' Attempts to Filter Content. Alkasir, a Case Study," in *SPIDER ICT4D* Series No. 3 (2010), 71–93.

6. Ronald J. Deibert et al., eds., *Access Denied: The Practice and Policy of Global Internet Filtering* (Cambridge: MIT Press, 2008).

7. In addition to my account here, Alkasir has also received extensive media coverage. See, for example, Magda Abu-Fadel, "Yemeni Develops Program to Skirt State's Web Bans, Gain Access to His News Portal," *Huffington Post*, 30 May 2009; available at:*www.huffingtonpost.com/magda-abufadil/yemeni-develops-program-t_b_209321.html*; Duncan Geere, "TED Fellows: Walid Al-Saqaf," *WIRED-UK*, 3 September 2010; available at *www.wired.co.uk/news/ted/2010-09/03/ted-fellows-walid-al-saqaf*; and Aleks Krotoski, "The Internet's Cyberradicals: Heroes of the Web Changing the World," *Guardian*, 28 November 2010; available at *www.guardian.co.uk/technology/2010/nov/28/internet-radicals-world-wide-web*.

8. Walid Al-Saqaf, "Unstoppable Trends: The Impact, Role, and Ideology of Yemeni News Websites," Masters dissertation, May 2008.

9. Doug Bernard, "Opening the Middle East Internet," *Voice of America*, 1 July 2010; available at *www.voanews.com/english/news/middle-east/Opening-the-Middle-East-Internet-97310199.html*.

10. Larry Diamond, "Why Are There No Arab Democracies?" *Journal of Democracy* 21 (January 2010): 93–104.

11. For more technical details, visit Alkasir's official website, *alkasir.com*.

12. This is the date on which Alkasir version 1.2.457 (stable) was released.

13. The number of Alkasir users does not necessarily mean the number of unique individuals using the program. It only measures the unique installations of the program. It may well be that the same program is used by more than one individual or that one individual may have multiple accounts (e.g., one at home and another at work).

14. Deibert et al., eds., *Access Denied*.

15. Naomi Sakr, "The Impact of Media Laws on Arab Digital and Print Content," background paper, *Arab Knowledge Report 2009: Towards Productive Intercommunication for Knowledge*, Mohammed Ben Rashid Al Maktoum Foundation in collaboration with UNDP, 2009, available at *www.mbrfoundation.ae/English/Documents/AKR-2009-En/AKR-English.pdf*.

16. The complete list of URLs blocked in different countries is available publicly online at *http://alkasir.com/map*.

17. Some countries appear less aggressive than others, blocking only websites that

meet a particular criteria (e.g., contain dissident content, reports criticizing the head of state, etc.), while other countries have an automated censorship solution (like Websense, Surf Nanny) that can block based on keywords in the URL or metadata.

18. Al-Saqaf, "Internet Censorship Challenged."

19. Global Internet Freedom Consortium, "New Technologies Battle and Defeat Internet Censorship" (2007); available at *www.internetfreedom.org/archive/Defeat_Internet_Censorship_White_Paper.pdf.*

20. The survey still exists online at *https://alkasir.com/doc/survey.*

21. Specifically, 2,600 of the 3,522 surveys had been fully completed by that date.

22. Jill Dougherty, "Digital Activists Skirt Roadblocks," CNN.com, 16 February 2011; available at *www.cnn.com/video/#/video/tech/2011/02/16/dougherty.digital.activism.cnn?iref=allsearch.*

23. A particular reported URL is initially approved if it is found to satisfy the policy at *http://alkasir.com/policy,* but it may require more effort to specify in which category it belongs.

10

SOCIAL MEDIA, DISSENT, AND IRAN'S GREEN MOVEMENT

Mehdi Yahyanejad and Elham Gheytanchi

Mehdi Yahyanejad *is the founder of the website* Balatarin.com, *the most popular social news website in Persian, with more than a million and a half viewers per month. He received his Ph.D. in Physics from MIT, with expertise in data mining and statistical methods.* **Elham Gheytanchi** *teaches sociology at Santa Monica College. Her research interests are the women's movement, social movements, and the use of information and communication technologies in Iran.*

Iran's 12 June 2009 presidential election—the tenth in the history of that country's Islamic Republic—was supposed to show the democratic face of the Iranian political system. Instead, the events following that disputed (and widely viewed as fraudulent) balloting inspired Iranians, and later many others across the broader Middle East, to rise up against their nondemocratic governments. The Iranian regime's suppression of dissent was then, and remains now, largely ineffective because of people's determination and creativity as well as their access to social media, the Internet, and mobile phones.

This chapter demonstrates that while the Iranian government has continued to press its censorship efforts, the Green Movement's activists have succeeded in using the Internet and social media to their advantage. As Larry Diamond correctly notes, "[I]t is not technology, but people, organizations, and governments that will determine who prevails."[1]

Some skeptics argue, however, that social media enable autocratic governments such as Iran's more than such media empower protesters. Malcolm Gladwell[2] and Evgeny Morozov[3] have stated that the Iranian government stays on top of technological advances and is therefore able to effectively block protests that originate online. But the protests are continuing. The existence of a grassroots civil rights social movement, born after the contested Iranian election of 2009, shows that widespread discontent and social mobilization will continue in spite of, and perhaps due to, the government's virtual and physical suppression.

This is not to say that censorship is completely ineffective. Starting on 25 January 2009 and continuing until a few months past the first anniversary of the disputed election, the Iranian government intensified its filtering and cyberattacks. Online suppression and surveillance by the government have indeed made it difficult to sustain the movement in Iran, but have failed to eradicate discontent or dismantle the loose networks of activists who carry on their nonviolent social struggle.

In their most recent book, *Access Contested*, and in the pages of the present volume, Ronald Deibert and Rafal Rohozinski describe three generations of censorship techniques used by governments to crush dissent.[4] The first of these involves tactics such as basic Internet filtering and monitoring. The more sophisticated second-generation techniques use legal pretexts for censoring website content or shutting down sites altogether, while third-generation techniques include surveilling citizens' online activities and other, more direct means of silencing dissenters.[5] In the aftermath of the presidential election, the Iranian government utilized such techniques in its attempt to censor the Internet.

Authoritarian political systems, including those of China and Iran, monitor the Internet to control the flow of news and information. As Rebecca MacKinnon explains, "When an authoritarian regime embraces and adjusts to the inevitable changes brought by digital communications technologies the result is 'networked authoritarianism.'" In this networked authoritarianism, ruling elites use the Internet for their own purposes while suppressing dissent through the same channels. While it is increasingly hard to sustain an opposition movement in the face of networked authoritarianism, Iran, like China, has a vibrant blogosphere despite government controls, arrests, and intimidation.[6]

In the months following the election, the Iranian government banned foreign journalists from reporting in Iran, arrested and jailed many Iranian journalists, and slowly enforced the monitoring of citizens' online activism. Even as the government was forcing foreign journalists to exit the country, however, social media were increasingly providing tools for reporting the news quickly and anonymously. Thus citizen journalism via social media made it possible for news to flow from Iran despite government censorship of the Internet and bans on foreign-media coverage.

This was possible because social media were relatively familiar and had been used by Iranians before the Green Movement. The Persian-speaking blogosphere helped to create the discourse for democracy, pluralism, and tolerance in the years prior to the election. Furthermore, reformist presidential candidates Mir Hosein Musavi and Mehdi Karrubi and their supporters used the Internet and social media throughout their campaigns. These campaigns took place on blogs, political websites, Balatarin (a Persian-language social news aggregator and networking site), microblogging sites like Twitter, and other social-networking sites such as Facebook, which Iranian youth used in great numbers early in 2009, months before the presidential election.[7]

While new social media played an undeniable role in the Green Movement, an unjustified emphasis has been placed on Twitter's role.[8] The international media and academia focused on Twitter because of three main factors. First, the content on Twitter was primarily in English and easily used for text analysis without a need for Persian-to-English translation. Second, Twitter has a powerful Application Programming Interface, which makes it easy to measure different types of activities such as the number of postings and re-tweets. Finally, the U.S. State Department's call for delaying Twitter's maintenance shutdown in June 2009 resulted in an exaggeration of the role that it performed.

A greater emphasis should be placed on the role of blogs. Blogs changed the political culture in Iran by allowing views that oppose the ruling party to appear online. This is important because, as Philip Howard has shown in his study of information and communications technologies (ICTs) in Muslim countries, "the route to democratization is a digital one."[9] In repressed societies, the online discussion of social, political, and cultural issues appears crucial to the success of social movements and transitions to democracy. As Howard states, "for democratic transitions, having a comparatively active online civil society is the most important ingredient of both necessary and sufficient solutions."[10] Through social media sites such as Balatarin, the Iranian online community can engage in debates, discuss political or social issues, and plan strategies for the accommodation of government supporters willing to compromise.

Iranian activists have used new social media to mobilize protest, expose injustice, inform fellow citizens as well as the world at large of human-rights violations, and facilitate transnational ties within the Iranian diaspora. This study of postelection social protest shows that, despite government control and censorship of social media, discontent grew and social mobilization for change occurred.

Furthermore, ICTs such as Balatarin have improved political rights in Iran and should thus be defined as "liberation technology." As Larry Diamond explains, liberation technology "means essentially the modern, interrelated forms of digital ICTs—the computer, the Internet, the mobile phone, and countless innovative applications for them, including 'new social media' such as Facebook and Twitter." In Iran, the Balatarin social-networking site, along with satellite-TV programs that echoed the voice of the Balatarin community, acted as "liberation technologies" because they allowed for the rapid and uncensored broadcasting of news and information. As the Green Movement's experience demonstrates, however, social media do have their limitations. Such media can become effective tools for spreading news and mobilizing people,

TABLE 1—THE FUNCTION OF DIFFERENT INTERNET PLATFORMS
DURING AND AFTER THE IRANIAN PRESIDENTIAL ELECTION

Internet Platform	Description
YouTube	YouTube was the most popular of the video websites. It helped to provide evidence for the demonstrations and to expose police brutality and killings. YouTube highlighted the Iranian protests on the front page of its CitizenTube during the demonstrations.
Facebook	The unblocking of Facebook by Iranian authorities before the election spurred rapid user growth and political activism. Facebook allowed information to flow outside of Iran and back.
Twitter	Twitter had a small following in Iran. However, a handful of Iranian activists and journalists tweeted in English which allowed foreigners and reporters to track events in Iran.
Political campaign websites inside Iran	Political websites based in Iran were influential because they provided content for discussion and sharing. GhalamNews and Kaleme, the official websites of Musavi, were reference points for many Musavi supporters. MowjSevvom, the campaign website of reformist candidates, brought former president Khatami back to the political scene as the presidential campaigns were launched. One of their petitions was signed by 450,000 people.
Blogs and Balatarin √	Blogs were a source of new ideas for the movement and also helped to expose human-rights violations by the Iranian government during and after the election. A number of nonviolent actions originated from the postings or discussions taking place on Balatarin.
Mowj Camp √	The Mowj Camp website became very popular after the Iranian presidential election. It was a source of daily news and ideas for the Green Movement. Several key members of Mowj Camp were arrested in Iran and the website shut down its operations. The Iranian authorities waged massive cyberattacks on Mowj Camp and other dissenting websites in the aftermath of the presidential election.[1]
News websites outside of Iran	A number of news websites based outside of Iran, such as Roozonline, Gooya news, Radio Farda, played an important role in exposing human-rights violations and compiling daily updates.
Blogs of *New York Times, CNN, Guardian,* and *Huffington Post*	Blogs of major newspapers allowed quick reporting on what was produced by citizen journalists.
Mailing groups and chain emails	Email lists provided the most reliable ways of distributing news or calls for protests in Iran.

[1]For a report on the intensity of cyberattacks on dissenting websites including Mowj Camp, see this report in Kaleme: *www.kaleme.com/1388/11/03/klm-9243/.*

but there is no substitute for essential strategic planning and leadership within a nonviolent social movement.

Balatarin, which means "highest" in Persian, is an online community website through which registered users can post their favorite website or blog and can vote to rank other links. Balatarin was launched in 2006 by Mehdi Yahyanejad and Aziz Ashofteh, and has since been the target of numerous cyberattacks. Balatarin allows the Iranian online community, as well as satellite-TV commentators, to aggregate blogs and websites from all over the Persian blogosphere into one public destination online.

TABLE 2—OTHER DIGITAL COMMUNICATION CHANNELS/DEVICES IN USE DURING AND AFTER THE IRANIAN PRESIDENTIAL ELECTION

Other Communication Channels	Description
Short Message Service (SMS)	SMS or texting by mobile phones is not reliant on the Internet. Consequently, it had a much wider reach in Iran and facilitated the sharing of information. The Iranian government blocked SMS frequently in the aftermath of the election.
Satellite broadcasts: BBC Persian and Voice of America (VOA)	Satellite-TV programs by the Iranian diaspora community, BBC Persian, and VOA have a wide audience in Iran. However, unlike Internet and social media forums, these broadcasts are not interactive.
Multi-Media Service (MMS)	Multimedia on mobile phones was transferred to the web, allowing the world to watch events in Iran despite the government ban on foreign journalists.

Balatarin was not the only social-media tool used by Iranians looking for information or for ways to mobilize. Facebook and a number of Iranian websites became places for the posting of news during and after the presidential campaign in Iran. A few journalists and activists used Twitter, but mainly to send information to the international audience. Due to slow Internet connections, YouTube content was not easy to watch in Iran, but the videos were broadcast through the BBC's Persian service, the Voice of America (VOA), and diaspora-run satellite-TV programs. Table 1 describes these and other Internet platforms and social media that were used during and after the Iranian presidential election of 2009.

Table 2 shows that Internet platforms were not the only liberation technology used during and after the presidential election. Other platforms, such as SMS, MMS, and satellite broadcasts, were important because they did not rely on the Internet.

Google searches for Balatarin, BBC Persian (the most influential news network outside Iran), and Fars News Agency (close to the Iranian Revolutionary Guard Corps or IRGC and supportive of President Mahmoud Ahmadinejad) all peaked following major events. The main peak occurred on Election Day, 12 June 2009. The second peak occurred later that year during the Ashura demonstrations of December 27. This correlation implies more demand for information prior to and during an event. Demand for news remained high for several days after each event.

Using the Internet as a safe haven for discussion is not a new tactic in Iran. After the Iranian government cracked down on the press in 2000, many Iranian journalists and activists found refuge on the Internet. Many Iranian activists living in Iran or abroad started blogging soon after that. In 2006, blogs in Persian ranked tenth among all languages worldwide in the number of posts placed online.[11] Journalists and activists used the Internet to carry on discussions that had begun offline during the era of relative press freedom under the presidency of Mohammad Khatami.[12] Whatever the medium of discussion, the content remained focused on how to build a civil society, how to work for reform, and how to promote the rule of law.

The Internet provided a forum where many new ideas and thoughts were openly shared and discussed. The dominant discussions on the web revolved around freedom of speech, freedom of assembly, and political reform. These discussions began to receive wider notice. More figures in the government and the official newspapers referred to blogs than had done so during the previous decade.[13]

During Ahmadinejad's first term (2005–2009), political dissent continued on the web. The Iranian government quickly moved to limit freedom of speech on the Internet. Yet as censorship increased, so did dissent,[14] and the government began using the Internet to counter this dissent for its own purposes, whether these involved spreading propaganda at home and abroad or concerned the internal affairs of Iran's factionalized government.

The June 2009 presidential election created an opportunity for the government to vaunt its tolerance of opposing views. The unblocking of Facebook and Twitter in January 2009 surprised many Iranian Internet users, and a large number soon signed up for these services. In less than a month, Facebook became the fifteenth-most-visited website in Iran.[15] Facebook soon became popular with reformists and social activists who were engaged in nonviolent social movements involving teachers, women, and students, to name just a few. Facebook registration grew rapidly until the government blocked it again in June 2009.

The Iranian government's relationship with social media has been tenuous and unpredictable. Authorities may at first have perceived social media as apolitical channels of communication worth leaving alone because they served to draw attention away from more blatantly political websites. Another explanation could be that the Iranian government viewed social media as a means of collecting information by, for instance, taking snapshots of users' profiles (Facebook did not provide encrypted access at the time).

Yet Facebook and Twitter were not the only players in this game. Reformist political activism began on a website called Mowj Sevvom (The Third Wave), which was devoted to calling on Khatami to run against Ahmadinejad. The site eventually collected more than 450,000 signatures on a petition urging the former president to run.[16] Khatami signed up for the presidential election and acknowledged the role of this online campaign in his decision to do so.[17] (Khatami would later leave the race when Musavi announced his candidacy.)

After Musavi's nomination the political campaign heated up quickly. Musavi made his campaign slogan "har Irani yek resaneh" ("every Iranian is a media outlet")[18] and called on his supporters to use the Internet to broadcast his message. Musavi maintained an active campaign website called Ghalam News. The government censored it on 21 June

2009, after it posted images of a defiant Musavi among cheerful and nonviolent demonstrators in Tehran the previous day. Not surprisingly, government-backed television and radio outlets had neglected to report on this demonstration.

The other reformist candidate, Mehdi Karrubi, was also active online. His party's newspaper, *Etemad-e Melli,* was regularly updated online. His campaign manager, Gholamhossein Karbaschi, created and maintained an active Twitter account. According to an advisor to one of the candidates, the two reformers even discussed who had enough "seniority points" on Balatarin.com and should be allowed to create the daily headline news for their campaign.[19]

Thus, the campaigns, protests, civil disobedience, and mass demonstrations following the controversial presidential election in 2009 were planned, reported, and widely discussed online through an already established social-media network.

Election Day: Crisis of Legitimacy

On election day, the Internet and social media provided "accountability technology," which, as Diamond explains, "enables citizens to report news, expose wrongdoings, express opinions, mobilize protest, monitor elections, scrutinize government, deepen participation, and expand the horizons of freedom." Just as Ushahidi, a free and open source-mapping software, enabled election accountability in Egypt and Sudan (as Patrick Meier shows in these pages), the Internet, social media, and mobile phones allowed Iranians to remain vigilant about widespread election fraud in their country.

The speed of events on 12 June 2009 caught everyone by surprise. The Iranian government disabled SMS across Iran. This damaged the efforts of political challengers to monitor the election. Musavi's camp had more than twenty-thousand observers in different polling locations. The observers were supposed to send SMS messages to Musavi's campaign headquarters in order to report results as well as any voting irregularities they might witness or suspect. Later, the Musavi campaign cited the interruption of SMS services as one of the primary pieces of evidence for fraud in the presidential election. The government, on the other hand, claimed that SMS service was interrupted in order to prevent illegal election-day campaigning by various candidates and their supporters.

The second event of the day was the widely reported attack by pro-regime militia forces on one of Musavi's main campaign offices, located in the North Tehran neighborhood of Gheytarieh. The Musavi campaign had set up a room there to handle webcasts of interviews with politicians and celebrities who were encouraging people to vote for Musavi. The hostile militia attacked the building to disrupt this broadcast. VahidOnline, an Iranian Internet celebrity who remained anonymous until that day, was present in the building. He witnessed the attack and posted his account of it on Twitter and his blog.[20] To document the severity of the situation for his online followers, he used his mobile phone to broadcast live footage of the attack to a video website called Qik.[21] Eight-thousand users viewed the video within hours. Meanwhile, the attackers made it to the fifth floor of the campaign headquarters and broke the broadcasting equipment. People in the building called the police, fought back, and were able to arrest four of the militia members.

The video of the attack on the Musavi headquarters was later shown on BBC Persian's evening news. It provided important evidence of the government's concerted effort to violently suppress the reformist candidates and their supporters. Videos, such as the one by VahidOnline, persuaded the public that a coup was underway. Supporters of the Iranian government claimed that security forces went to the headquarters in order to disable a broadcast that was violating campaign laws.[22] VahidOnline went into hiding and later escaped Iran. He posted an emotional statement on his blog entitled "The Crime of Being Online" before crossing the border with the help of smugglers.[23]

As the events of the day unfolded, the perception of a coup by government-backed militia strengthened. At 6:30 p.m. (all times are local), several hours before the polls closed, Fars News Agency, a website close to the IRGC, predicted that Ahmadinejad would win with 60 percent of the vote. The Iranian online community reacted with disbelief. The announcement from Fars News was posted to Balatarin with the altered title of "Is This Believable: Musavi 28 Percent!! Ahmadinejad 69 Percent? (The Biggest Fraud of the Century Has Begun)." This link was posted on Balatarin a few minutes after midnight, only three hours after polling stations had closed.[24] This posting marked the establishment of Balatarin as the hub of what would later be called the Green Movement.

The Case of Neda's Video

Government-backed militia killed Neda Agha-Soltan at 7:20 p.m. on 20 June 2009. The person who took the video of the unarmed young woman being shot down and swiftly bleeding to death on a Tehran street sent it to a contact outside Iran. That person posted it on Facebook at 8:53. The link to the video appeared on Balatarin at 9:45, and within fifteen minutes had received enough votes to be promoted to the front page. The video was subsequently posted to YouTube at 10:19. Thus in less than three hours the video of Neda's death had been seen by thousands of viewers worldwide. This video has come to represent the Iranian government's violent repression of its citizens' nonviolent protest movement.[25]

Thanks to widespread use of social media and mobile phones and to the advent of citizen journalism, street-level violations of human rights by Iranian officials could no longer be kept in the dark. An Iranian dissident whom I shall call S.E. said in an interview,

> In 1988, thousands of Iranians were killed in mass executions inside the Iranian prisons within a short period of time and to this day the prison massacre has received very little publicity. Now a good number of atrocities have been documented, and the Iranian government is forced to come up with explanations for them.

Frequently these explanations have been lies. For example, during a massive nonviolent demonstration in Tehran on 26 December 2009, a police truck drove over a protester several times.[26] The video was immediately posted on YouTube, Balatarin, and Facebook, and was broadcast in Iran via satellite TV. The government had to provide three different explanations for the scene.[27] Similarly, the government changed its story regarding Neda's fate many times.[28] At first, the government claimed that Neda had not died and that the video showed fake blood. Later, officials made and gave nationwide television airtime to a "documentary" that portrayed Neda as the victim of a female assassin who ran from the scene.

In Iran, television and radio broadcasting as well as major newspapers are under the direct supervision of the Supreme Leader. Iranian press and TV experienced short periods of relative freedom of expression during the revolutionary days in 1979 and again during Khatami's first term (1997–2001). Prior to the presidential election of 2009, satellite broadcasts by BBC Persian and VOA enjoyed a wide reach among the general public, but there is currently no such thing as free TV, radio, or press in Iran. Most of the newspapers in Iran are state-owned or belong to progovernment politicians.

Students, journalists, and activists use Persian-language websites as their main source of alternative news. Internet censorship, however, means that numerous visitors still look at government-owned sites such as Mehr and Fars News. For example, Fars News Agency has more visitors than Radio Farda, the Persian-language arm of Radio Free Europe funded by the U.S. government. Fars also draws more visitors than Gooya News, Balatarin, or the Green Movement's Jaras, all of which are independent websites based outside Iran.

Information from social media and citizen journalists continued to flow despite the government's repeated jamming and censorship. News from these independent sources affected mainstream media coverage. News networks such as CNN started broadcasting YouTube videos from Iran. BBC Persian and VOA also began using uploaded videos from YouTube in their reporting. One could argue that it gradually became difficult for these networks to prevent their airtime from becoming a platform for social action because many of their journalists were supporting the movement.

Nonviolent mass demonstrations on Quds Day (18 September 2009) demonstrated the interaction of satellite-TV programming with new social media in Iran. The last Friday in the month of Ramadan is called Quds (Jerusalem) Day and is used by the Iranian government for state-supported anti-Israel demonstrations. In mid-2009, after the violent government crackdowns on protests, regime officials believed that the demonstrations had been stamped out. This belief was wrong. Online activists called for Green Movement supporters to show up at the Quds Day demonstrations in order to protest the crackdown. The Green Movement activists called it "Quds Day, the Green Day of Iran."[29]

This mobilization had begun several weeks earlier and at first had drawn no support from opposition figures in Iran. Many of them had felt hesitant to use demonstrations against Israel as an occasion to protest against the Iranian government. Only as Quds Day drew near did Karrubi respond to the calls of online activists.[30] Musavi waited until the last day to announce his participation. The Quds Day demonstration, therefore, was born on the Internet.

Because the government disrupted Internet services on Quds Day, no YouTube videos of the demonstrations were available for several hours. This led BBC News to publish an erroneous article that stated,

> Reformist opponents of the controversially re-elected President Ahmadinejad seem to have been massively outnumbered by system loyalists eager to demonstrate their support for the president and his patron, the Supreme Leader, Ayatollah Ali Khamenei.[31]

In response, Balatarin users posted a link with the title "The weird claim of international media regarding the low number of the Green Movement activists in the demonstrations shows there is a need to send pictures." Online activists quickly posted links to pictures and new videos on YouTube to show the large demonstrations by Green Movement supporters.[32] The BBC website updated the article without even acknowledging the correction: "Thousands of opposition supporters have clashed with security forces during a government-sponsored rally in Tehran."[33] Thus did social media, acting as a channel for street-level reporting by viewers, force a major international news outlet to correct a story.

Far from having given up, the Green Movement protesters were present in large numbers and subverted the official demonstration.[34] In a way, the Quds Day protest transformed the Green Movement into a longer-lasting movement. Musavi was well aware of the significance of the demonstration. He called it a turning point for the movement.[35] In a speech after the demonstration, he acknowledged social media's role:

> Today, there has been a network created in the virtual space acting very efficiently when there isn't any other type of [independent] media available. The social groups acting within this virtual space are less venerable. Members of these groups have given dynamism to the movement, which has made us much more hopeful. There hasn't been any official call [by the leaders of the movement] for a demonstration on Quds Day, but we witnessed this great demonstration. This was at a time when there had been many, many threats in the past three months and many of the families were preventing their children from going [to the demonstrations]. This could have not have been achieved without this [virtual] network.[36]

The mass protest that followed the 19 December 2009 death (from natural causes) of reformist ayatollah Hussein-Ali Montazeri again demonstrated how activists could and did use social media, satellite TV, SMS, and word of mouth to mobilize. The idea for a major demonstration originated on a blog and was publicized through Balatarin. An anonymous blogger posted a link on Balatarin suggesting that protesters should gather in Tehran's Mohseni Square in order to mourn Montazeri publicly. In less than nine hours, about three-thousand people had shown up at the square.[37] It is important to note that the blogger in this case acted alone and grassroots groups implemented the idea without prior coordination with reformist leaders Karrubi or Musavi.[38]

New social media became so pervasive that they were able to utilize Iranian national TV as an unwilling platform. Take, for example, the symbolic "green" voting during the most-watched show on Iranian state TV, a live sports program called "90." During this program, people are asked a question and invited to vote for one of several answers—each one associated with a different color—via SMS. Early in January 2010, online activists asked people to send an SMS to the next program of "90" and choose the green option regardless of what the question was.[39] The idea for this nonviolent tactic spread quickly through SMS in Iran. The last choice was chosen because it was usually shown in green. During the program, the host changed the last option from green to yellow, but people still voted for the last option. More than 1.8 million people voted and 75 percent of them chose the third option (which was not the correct answer to the particular question being asked). This simple and fairly low-risk action proved to all those watching the program that at least a million Iranians were ready to show their dislike of the government by following the campaign of the Green Movement.[40]

Another effective protest that was entirely conducted online concerned political prisoner Majid Tavakoli, a student activist who had been arrested after giving a speech in which he criticized Iran's leaders. Fars News Agency claimed that Tavakoli was arrested when he tried

to escape wearing woman's clothing. It was evident this had been published to discredit Tavakoli.[41] Masih Alinejad, a journalist and blogger,
posted an article asking men to wear headscarves in solidarity with Majid and to protest the forced Islamic dress code for women in Iran.[42] This
was posted on Balatarin and became the most voted item of the day and
quickly spread to Facebook and other social networks.[43] More than 450
men took pictures of themselves with headscarves and posted them on
Facebook in support of Tavakoli.[44]

While many platforms had a significant impact on the Green Movement, some are ascribed a greater influence than they deserve. Many
people using the phrase "Twitter Revolution" imagined activists running in the streets of Tehran, coordinating demonstrations and tweeting
about their future plans. That never happened. As mentioned before,
there were not that many Twitter users in Iran, and at critical moments
the Iranian government disabled the SMS system on which Twitter relies. Other forms of social media were used in a number of ways to call
for action, but these calls were not always successful.

When Social Media Failed

When the large demonstrations on the eve of the Shia religious ceremo-
ny of Ashura (27 December 2009) were over, Green Movement activists
began discussing ideas for a similar mass demonstration on the upcoming
February 2010 anniversary of the 1979 revolution. The annual demonstration, organized by the state, starts from the main street in central Tehran
and leads to the large-Azadi (Freedom) Square. Ebrahim Nabavi, a well-
known Iranian satirist and blogger, suggested that the protestors hide their
green signs until arrival at Azadi Square. Once there, he suggested, they
would be able to take over the square and disrupt Ahmadinejad's speech.
Nabavi posted his idea on Jaras. The idea was debated online and most
commentators agreed with the plan. In a poll conducted a day before Ahmedinejad's scheduled speaking date of 11 February 2010, 80 percent of
visitors to Balatarin said they believed that this strategy would succeed.
Many of the Green Movement activists hoped that it would compare to the
day (21 December 1989) in Bucharest when a jeering crowd disrupted a
speech by Romanian dictator Nicolae Ceaușescu, who fled his capital and
soon thereafter fell from power.

When the day of action arrived, the Iranian government showed that it
had been preparing for this event. Security forces controlled the streets of
Tehran and citizens were afraid to leave their houses. Government supporters were bused into the main street leading to Azadi Square. Green Movement supporters who made it to the rally were too spread out and were
unable to find each other and coalesce. Security forces blocked all routes
leading to Azadi Square. Only an already selected group was allowed to
enter and was placed in the front row near Ahmadinejad's podium.

A satellite picture taken by Geo Eye during the rally showed exactly what had happened.[45] Azadi Square had been kept mainly empty. In the satellite image, one could make out a large number of buses that had carried government supporters to the main street leading to Azadi Square. The Trojan horse strategy had obviously failed. The Iranian government knew the intentions of the Green Movement supporters well in advance and prepared for them. In hindsight, many people realized the protest plan had been impractical and weak on crucial logistical details such as means for countering blocked roads and the government maneuvers. This failed demonstration was the Green Movement's last attempt (so far) at mounting a major street protest.

Clearly, new social media such as mobile phones, Balatarin, and satellite TV acted as "liberation technology" during and after the disputed 2009 presidential election in Iran. New media lowered the cost of political participation and protest, and proved crucial as the only channels through which large-scale demonstrations could be effectively coordinated down to specifics of date, time, and place. The risks of belonging to Balatarin, or a Facebook page connected with the Green Movement, or a mailing list of those receiving Jaras updates were acceptable to many who were seeking change. These loose, online affiliations have lowered the costs of mobilization and membership for progressive social movements. Though social media can widen the grassroots base of social movements, such media (with their open, horizontal nature) can also breed confusion when there is a need to deal with complex issues and tactics that require discipline, strategy, and a degree of central leadership. While social media worked effectively to mobilize protesters in a grassroots nonviolent social movement, better organization and planning will be needed if the Green Movement is to become victorious.

NOTES

1. Larry Diamond, "Liberation Technology," *Journal of Democracy* 21 (July 2010): 82.

2. Malcolm Gladwell, "Small Change: Why the Revolution Will Not Be Tweeted," *New Yorker,* 4 October 2010, available at *www.newyorker.com/reporting/2010/10/04/101004fa_fact_gladwell.*

3. Evgeny Morozov, *The Net Delusion: The Dark Side of Internet Freedom* (New York: PublicAffairs, 2011).

4. Ronald Deibert et al., *Access Contested: Security, Identity, and Resistance in Asian Cyberspace* (Cambridge: MIT Press, 2012).

5. Ronald Deibert and Rafal Rohozinski, "Beyond Denial: Introducing Next-Generation Information Access Controls," in Deibert et al., eds., *Access Controlled: The Shaping of Power, Rights, and Rule in Cyberspace* (Cambridge: MIT Press, 2010), 6.

6. Nasrin Alavi, *We Are Iran* (London: Portobello, 2006); Guobin Yang, *The Power of the Internet in China: Citizen Activism Online* (New York: Columbia University Press,

2009); John Kelly and Bruce Etling, "Mapping Iran's Online Public: Politics and Culture in the Persian Blogosphere," Berkman Center for Internet and Society, Harvard University, 5 April 2008.

7. Omid Habibinia, "Who Is Afraid of Facebook?" 3 September 2009, available at *http://riseoftheiranianpeople.com/2009/09/03/who-is-afraid-of-facebook*.

8. Zicong Zhou et al., "Information Resonance on Twitter: Watching Iran," First Workshop on Social Media Analytics (SOMA '10), Washington, D.C., 25 July 2010. See *http://snap.stanford.edu/soma2010/papers/soma2010_17.pdf*.

9. Philip N. Howard, *The Digital Origins of Dictatorship and Democracy: Information, Technology, and Political Islam* (New York: Oxford University Press, 2010), 201.

10. Howard, *Digital Origins*, 183.

11. "Technocrati: Persian Among the Top Ten Languages in Blogsphere," 9 November 2006, available at *www.bbc.co.uk/blogs/persian/2006/11/post_132.html*.

12. Babak Rahimi, "Cyberdissent: The Internet in Revolutionary Iran," *Middle East Review of International Affairs* 7 (September 2003), available at *http://meria.idc.ac.il/JOURNAL/2003/issue3/rahimi.pdf*.

13. Aiden Duffy and Philip N. Howard, "Iran's Political Parties Link to Persian Blogosphere More than News Sources," Project on Information Technology and Political Islam, University of Washington–Seattle, 2010.

14. Elham Gheytanchi and Babak Rahimi, "Iran's Reformists and Activists: Internet Exploiters," *Middle East Policy Journal* 15 (Spring 2008): 45–59.

15. For a Persian-language discussion of Facebook's popularity in Iran, see *http://mhmazidi2.wordpress.com/2009/02/26/fa-to-15*.

16. The list of those who signed the petition urging Khatami to run is available at *www.mowj.ir/PatitionList.php*.

17. "Khatami talking to Mowj reporter," *www.youtube.com/watch?v=DiTq5Tjng8Y*.

18. See *www.kaleme.com/1388/11/13/klm-10327*.

19. Phone interview with authors, May 2009.

20. Exclusive phone interview with VahidOnline (unpublished), January 2010.

21. See *http://qik.com/vahidonline*.

22. See *www.farsnews.com/newstext.php?nn=8806231161*.

23. "The Crime of Being Online," *http://vahid-online.net/1388/05/20/be-online*.

24. See *http://balatarin.com/permlink/2009/6/12/1617273*.

25. Metter Mortensen, "When Citizen Photojournalism Sets the News Agenda: Neda Agha Soltan as a Web 2.0 Icon of Post-Election Unrest in Iran," *Global Media and Communication Journal* 7 (April 2011): 14–16.

26. "A Protester Is Run Over by an Armored Truck," *www.youtube.com/watch?v=J2d3M203aAg&feature=related*.

27. "Tehran Police Chief Doesn't Admit That the Truck Belonged to Security Forces," *www.youtube.com/watch?v=YZbDU8KBE98&feature=related.*

28. "Iran Ambassador Suggests CIA Could Have Killed Neda Agha-Soltan," *http://latimesblogs.latimes.com/washington/2009/06/neda-cia-cnn-killing.html*; "Zarghami Claims That Neda's Death Is Fake," *http://khabaronline.ir/news-11915.aspx*; Press TV documentary of Neda's death, *www.youtube.com/watch?v=Shp7HE2YA_c&skipcontrinter=1.*

29. See *www.kaleme.com/1389/06/11/klm-30708.*

30. "Karoubi: You Will See the Power of People on the Quds Day," *www.radiofarda.com/content/f3_karoubi_protest_Iran_Quds_rally_green_movement_postelection/1810580.html.*

31. This was deleted from the website and replaced with the content in endnote 34 without acknowledging the correction.

32. Robert F. Worth, "Despite Warning, Thousands Rally in Iran," *New York Times,* 18 September 2009.

33. "Clashes Erupt at Iran Mass Rally," 18 September 2009, *http://news.bbc.co.uk/1/hi/world/middle_east/8262273.stm.*

34. "Iran Protests on Quds Day," *Guardian,* 17 September 2009, *www.guardian.co.uk/news/blog/2009/sep/17/iran-protests-quds-day.*

35. "Mousavi Calls the Quds Demonstration a Turning Point," *www.bbc.co.uk/persian/iran/2009/09/090928_si_mousavi_Qudsday.shtml.*

36. Musavi's speech following Quds Day is available at *www.parlemannews.ir/?n=4128.*

37. "Video of the Gathering After the Death of the Dissident Ayatollah Montazeri," *www.youtube.com/watch?v=KLNteL_V6QY.*

38. "The Call for Gathering to Commemorate the Dissident Ayatollah's Death," *http://balatarin.com/topic/2009/12/20/1003993.*

39. "The Idea of Sending One Million Green SMS to the Program *90*," *http://balatarin.com/permlink/2009/9/21/1766073.*

40. "The Result of the SMSs Sent to the Sport Program," *www.youtube.com/watch?v=LrfigqKfuD0&feature=player_embedded.*

41. Robert Tait, "Iran Regime Depicts Male Student in Chador as Shaming Tactic," *Guardian,* 11 December 2009, *www.guardian.co.uk/world/2009/dec/11/iran-regime-male-student-chador.*

42. "The Story of Majid Tavakoli Is the Story of Humilation of the Women in My Country," *http://masihalinejad.com/?p=953.*

43. "The Green Movement Wear Headscarf in a Symbolic Gesture," *http://balatarin.com/permlink/2009/12/9/1867237.*

44. Amy Kellogg, "Iran's Veil Campaign," 11 December 2009, *http://liveshots.blogs.foxnews.com/2009/12/11/veil-campaign.*

45. Satellite picture of the progovernment rally, 11 February 2010, available at *www.geoeye.com/CorpSite/gallery/detail.aspx?iid=294&gid=20.*

IV

Policy Recommendations

11

CHALLENGES FOR
INTERNATIONAL POLICY

Daniel Calingaert

Daniel Calingaert *is vice-president for policy and external relations at Freedom House, which receives funding from the State Department, Google, and other sources to promote Internet freedom. He also teaches in Georgetown University's M.A. Program in Democracy and Governance and at Johns Hopkins University's School of Advanced International Studies.*

In her groundbreaking speech on Internet freedom in January 2010, U.S. secretary of state Hillary Clinton hailed the "spread of information networks" as the "new nervous system for our planet" and highlighted their importance in "making governments more accountable." She warned, however, that they "can be harnessed for good or for ill." In May 2010, then–French foreign minister Bernard Kouchner echoed her assessment of the promise of and challenges for information and communications technologies. He praised the Internet as "the most fantastic means of breaking down walls that close us off from one another" but expressed concern about the "alarming rate" at which "the number of countries that censor the Internet and monitor Web users is increasing." He, like Clinton, called for engagement in the "battle of ideas" between "the advocates of a universal and open Internet" and "those who want to transform the Internet into a multitude of closed-off spaces that serve the purposes of repressive regimes."[1]

U.S. and European governments are engaging in the struggle for on-line freedom of expression. U.S. support for Internet freedom now is clearly articulated, identified as a priority in U.S. foreign policy, and backed by significant resources. European governments have defined their approaches and begun to make contributions to Internet freedom. Their methods have both strong similarities to U.S. policy and distinctive elements.

Thus, this is an opportune moment to assess the direction and initial impact of government policy on Internet freedom and to explore effective ways to move forward. This chapter begins with a summary of the main challenges to Internet freedom globally and then examines the U.S. government's record to date in promoting Internet freedom, particularly in balancing Internet freedom against competing political, security, and economic interests. It then compares the U.S. government's approach to Internet freedom to that of European counterparts, and it concludes with recommendations for improving U.S. policy.

The State of Global Internet Freedom

Internet freedom depends to a large extent on how much the United States and other democratic countries support it.[2] Information technologies and digital media, if left alone, have great potential to liberate. They give citizens access to a wide range of information that is unavailable in traditional media. Web 2.0 applications allow just about anyone with a computer, Internet connection, and fairly rudimentary skills to become a publisher and distributor of information. News and political commentary are no longer in the exclusive domain of large media houses, whose editors decide what is fit to print and what is not. Such commentary is, rather, disseminated by myriad sources, often through individual initiative.

Moreover, Web 2.0 applications provide powerful tools to organize citizen activism. They have been used to organize mass protests across the Middle East and North Africa, to spread news of the protests and the crackdowns that ensued, and to disseminate videos to document the scale of the protests and the brutality of the crackdowns. The power of digital media was particularly evident in Syria, which banned foreign journalists—including regional Arab media—during the unrest and arrested thousands of activists and protestors. Antiregime activists used Twitter to alert the international media about the situation in Syria, posted videos on YouTube of security forces shooting at demonstrators, and set up Facebook groups to coordinate slogans and plans among separate groups of activists in different cities. Web 2.0 applications therefore may merit the moniker of "liberation technology."

The liberating effects of information and communications technologies (ICTs) are well known to authoritarian regimes, and, for precisely this reason, these regimes have introduced extensive controls over digital media. The most sophisticated authoritarian regimes, such as China and Iran, have become highly adept at controlling the Internet. They maintain pervasive, multilayered systems of censorship and surveillance, which stifle independent expression and online dissent and obstruct the ability of political opposition to organize through social media. They manage to promote Internet access for the purposes of stimulating technological innovation and economic growth but, at the same time, they greatly restrict the use of ICTs to challenge their hold on power.

Internet restrictions are mounting around the world. As documented in Freedom House's 2011 report on *Freedom on the Net*, threats to on-line freedom are growing and are increasingly diverse.[3] Of the fifteen countries covered in the 2009 edition of *Freedom on the Net*, nine registered declines in the subsequent two years. Several countries that previously had few Internet controls, including Jordan and Venezuela, have begun to censor political content and violate Internet-user rights. Jordan introduced a law in August 2010 with heavy fines for posting defamatory comments online and harsh penalties—fines and jail sentences—for posting previously unpublished information on Jordan's national security, public order, or economy. In December 2010, Venezuela enacted laws that increased state control over telecommunications networks and laid the foundation for requiring website managers and service providers to censor user comments. Azerbaijan's Ministry of National Security has proposed amendments to the criminal code to make the act of spreading "misinformation" a "cybercrime."[4] In other countries, the trend toward greater control has accelerated dramatically. Pakistan set up an Inter-Ministerial Committee for the Evaluation of Websites in mid-2010 to flag sites for blocking based on vaguely defined offenses against the state or religion. In Thailand, court orders have blocked tens of thousands of websites, and dozens of citizens have faced charges for views they expressed online, particularly for views that were critical of the monarchy.

The variety of controls exercised over the flow of online information includes bans on social-media applications, denial of Internet access, intermediary liability for service providers, online surveillance, and digital attacks. These controls have intensified over the past two years. Blocks on social-media applications have spread widely. In 12 of the 37 countries analyzed in the 2011 report of *Freedom on the Net*, authorities consistently or temporarily imposed total bans on YouTube, Facebook, Twitter, and similar applications.

Governments have also used their control of Internet infrastructure to facilitate censorship and, at times of political unrest, to cut off access to the Internet. Chinese authorities severed all online connections to the Xinjiang region from July 2009 to May 2010, while security forces carried out mass arrests in the wake of ethnic violence there. Egypt shut off Internet access for the entire country for five days in late January 2011 amidst mass protests calling on then-president Hosni Mubarak to step down. In March 2011, the regime of Muammar Qadhafi cut off virtually all Internet access in Libya in an attempt to stem the antiregime uprising. Moreover, some countries are trying to insulate their citizens from the global Internet. Iran is taking steps toward the creation of a national internet to disconnect Iranian users from the rest of the world.[5] North Korea and Cuba already maintain national internets and allow only a very small segment of their population to access the Internet.

Hosting companies and service providers are increasingly held liable for the online activities of their users. In Thailand, the host of an online platform is facing criminal charges over reader comments that were critical of the monarchy. In Vietnam and Venezuela, some webmasters and bloggers have disabled the comment feature on their sites to avoid potential liability. In Belarus, a government decree went into effect in mid-2010 requiring Internet service providers to register with the Ministry of Communications, provide technical details about online resources and systems, and identify all devices that are being used to connect to the Internet. These provisions are designed to give the Belarusian government greater capacity to monitor online activities. Belarus also introduced requirements for Internet cafés to check the identity of users and keep a record of their web searches.

Online surveillance appears to have grown more extensive over the past two years. The Iranian government used intercepted online communications, including activities on Facebook and the Persian-language social-media site Balatarin, to prosecute activists involved in protests against the fraudulent 2009 presidential election. Many arrested activists reported that interrogators confronted them with copies of their e-mails, demanded the passwords to their Facebook accounts, and questioned them about individuals on their friends list.[6]

Western companies have come under pressure to facilitate government interception of digital communications. The governments of Saudi Arabia and India pressured BlackBerry maker Research In Motion (RIM) to grant them access to data of BlackBerry users, and the United Arab Emirates made similar demands on RIM.[7]

Digital attacks against human-rights and democracy activists have become widespread. They have crippled independent or opposition websites and disrupted communication and collaboration among activists. Sophisticated and extensive cyberattacks have originated from China. These included denial-of-service attacks on domestic and overseas human-rights groups, e-mail messages to foreign journalists containing malicious software capable of monitoring the recipient's computer, and a cyberespionage network that extended to 103 countries to spy on the Tibetan government-in-exile. In Iran, during the mass demonstrations that followed the June 2009 presidential election, denial-of-service attacks disabled many opposition news sites, and the Iranian Cyber Army, which operates under the command of the Iranian Revolutionary Guard Corps, hacked several other prominent websites. In Belarus, to stifle protests against the fraudulent December 2010 elections, denial-of-service attacks slowed down connections to opposition websites or rendered them inaccessible. Moreover, the country's largest Internet service provider, the state-owned Belpak, redirected users from independent media sites to nearly identical clones that provided misleading information, such as the incorrect location of a planned opposition rally. Digital attacks on websites or blogs that are critical of the government have also taken place in several countries rated Partly Free on Internet freedom by Freedom House, including Kazakhstan, Malaysia, and Russia.

The trend toward growing restrictions on the Internet looks likely to continue. China has outlined plans to intensify its Internet restrictions, even though they already are among the most extensive in the world. These plans were laid out in the April 2010 report on the development of China's Internet by Wang Chen, director of the Communist Party's External Propaganda Department and director of the Information Office of China's State Council. He lamented "the difficulty in the supervision of information content security" with the proliferation of audio, video, and other multimedia content and the expansion of mobile Internet services. He proposed to respond by intensifying "efforts to guide public opinion online" and strengthening "efforts to manage domain names, IP addresses, registration and record-filing, and connection services." Chen mentioned plans to introduce administrative procedures "for managing mobile media and implementing real name registration of cell phones." Additional measures will be taken to "make the Internet real name system a reality as soon as possible" and to "gradually apply the real name registration system to online interactive processes." These plans aim, as Chen explained, to "perfect our system to monitor harmful information on the Internet, and strengthen the blocking of harmful information from outside China."[8]

U.S. Policy

In response to the growing threats to Internet freedom around the world, the U.S. government has presented a clear set of policy goals for promoting freedom of expression online, undertaken diplomatic efforts to promote these goals, and allocated substantial resources to counteract restrictions on the Internet. The goals are defined in broad terms. In her January 2010 speech, Secretary Clinton called for the United States "to help ensure the free exchange of ideas" online and promised to commit "the diplomatic, economic, and technological resources necessary to advance these freedoms." President Barack Obama, in his address to the United Nations General Assembly in September 2010, declared that the United States must "leverage new technologies so that we can strengthen the foundation of freedom." The U.S. government, as Clinton explained in February 2011, promotes "an Internet where people's rights are protected" and that is "interoperable all over the world" and "secure enough to hold people's trust."[9]

The Obama administration has elevated Internet freedom to a priority in U.S. foreign policy and a key component of its human-rights agenda. It has articulated clear and forceful policy positions on Internet freedom and pushed vigorously at the UN and elsewhere for online freedom of expression. Rather than develop new international human-rights instruments for the Internet, the Obama administration has used existing internationally recognized human-rights principles, such as Article 19 of the Universal Declaration of Human Rights, to promote Internet freedom. Article 19 upholds the "right to freedom of opinion and expression," including the freedom "to seek, receive, and impart information and ideas through any media and regardless of frontiers."

Diplomatic efforts at multiple levels seek to promote online freedom of expression. The U.S. government has resisted attempts to place Internet governance under the UN, specifically the International Telecommunication Union, where authoritarian regimes may have greater ability to introduce new controls.[10] The United States instead supports the multistakeholder bodies that currently govern the Internet, such as the Internet Corporation for Assigned Names and Numbers (ICANN). The State Department addresses Internet-freedom issues in its engagement with foreign governments, for instance in its human-rights dialogues with China. According to Secretary Clinton, these issues have become "part of the daily work" of diplomats around the world, who are charged with "monitoring and responding to threats to Internet freedom."[11]

The commitment of resources to promote Internet freedom is substantial. The State Department has spent US$50 million since 2008 on a range of Internet-freedom programs, and Congress allocated another $20 million to the department for 2012. These programs have supported technologies to circumvent online censorship, secure tools for mobile phones, efforts to reintroduce blocked content to users behind a firewall, and digital security training for activists.[12] In addition, the Broadcasting Board of Governors has made significant investments in censorship circumvention,[13] and the U.S. Agency for International Development has provided substantial support for digital media in repressive environments and for efforts by civil society to enhance the use of information and communications technologies for civic activism.[14]

Secretary Clinton explains that the State Department programs take a "venture capital-style approach." They seek to spur technological innovation. They support a portfolio of technologies, tools, and training that aims to respond to needs identified by digital activists in repressive environments, including the growing needs for mobile-phone technologies and digital security. Moreover, these programs support multiple tools, so that "if repressive governments figure out how to target one, others are available."[15]

U.S. government funding for Internet-freedom programs is needed and justified. It addresses challenges and opportunities that businesses and private donors cannot adequately meet on their own. It builds on decades of U.S. government experience in supporting programs that promote human rights and democracy around the world.

The State Department has taken a sophisticated approach to supporting Internet-freedom programs. It focuses on programs that are unlikely to attract private donations, such as technologies for users in repressive environments who rely on web applications that are free of charge. These programs address a range of priorities identified by digital activists in repressive environments. They aim to stimulate innovation in censorship-circumvention and privacy-protection tools, to expand access to these tools in Internet-restrictive environments, to assist human-rights defenders and civic activists in applying new technology to their work, and to support the activists' efforts in challenging the limits to online freedom of expression imposed by repressive governments. The State Department has come under criticism, however, for taking more than a year to spend $30 million in 2010 funds appropriated for Internet-freedom programs,[16] and for supporting Haystack, a tool that purported to give Internet users in Iran the ability to bypass firewalls and communicate securely but that in fact had serious security flaws.[17] The experience with Haystack underscores the risks associated with State Department support for technological innovation.

Competing U.S. Interests

Current U.S. policy is inadequate to stem—let alone reverse—the trend of declining Internet freedom. Growing restrictions by repressive regimes are outpacing U.S. efforts to enable information and communications technologies to expand the space for free expression. Despite U.S. diplomatic initiatives, responses to violations of Internet users' rights, and Internet-freedom programs, authoritarian regimes are still increasing their controls over the Internet. Moreover, when Internet freedom competes with other U.S. political, security, or economic interests, the other interests tend to take precedence over Internet freedom.

In pushing back against Internet-related repression, the U.S. government's record is quite strong overall, but weak in certain respects. The State Department is increasingly standing up for bloggers and cyberdissidents who come under pressure from repressive regimes. It invited several leading bloggers to attend Secretary Clinton's Internet freedom speech in January 2010, including Bassem Samir, who was a thorn in the side of the Mubarak regime in Egypt. The U.S. government has criticized the arrests of prominent bloggers and cyberdissidents, such as Bahraini blogger Mahmood al-Yousif. It is, however, less vocal in its criticism of Internet-freedom violations committed by certain foreign governments, such as Pakistan and Thailand.

U.S. support for cyberactivists has encountered skepticism in some quarters, particularly in the Middle East and North Africa, because it coincided with U.S. collaboration on counterterrorism and other issues with authoritarian regimes. It put the United States in the awkward position of supporting both authoritarian regimes and the citizens they oppress or, in the words of journalist Rami Khouri, of "feeding both the jailer and the prisoner." Khouri derided the U.S. government for supporting "political activism by young Arabs while it simultaneously provides funds and guns that help cement the power of the very same Arab governments the young social and political activists target for change."[18]

In the face of restrictive Internet laws and policies of authoritarian governments, the U.S. response has been mixed. U.S. officials have condemned such policies at times, for instance when Secretary Clinton mentioned China's Internet censorship in her February 2011 speech. At other times, officials were largely silent when new Internet restrictions were introduced, for instance when Saudi Arabia introduced a requirement in early 2011 for online media sites (including blogs) to obtain a license to operate.

Moreover, to counteract Internet censorship, the Obama administration has offered incentives for repressive regimes to allow greater freedom online but has not spelled out any consequences if these regimes maintain their censorship. When State Department officials led a delegation of representatives from U.S. technology companies to Syria in June 2010, the U.S. officials suggested to Syrian president Bashar al-Assad that if he stopped blocking social-media sites such as Facebook and YouTube, U.S. investment was likely to follow. They neglected, however, to publicly criticize Syria's Internet censorship or to meet with independent bloggers or cyberdissidents.

In addition to political interests, security concerns compete with the U.S. interest in Internet freedom. The open nature of the Internet creates vulnerabilities for U.S. military assets and critical infrastructure. National security officials seek to reduce these vulnerabilities and, in the process, have proposed policies that may put the open nature of the Internet at risk. The commander of U.S. Cyber Command, General Keith Alexander, has called for the creation of a separate secure computer network for government agencies and critical industries to insulate them from digital attacks.

While such a network would enhance the security of government and industry computer networks, it might also accelerate a move toward greater fracturing of the Internet, whereby more countries seek to wall off parts of their Internet. General Alexander has also requested authority to carry out offensive operations on the Internet, including preemptive cyberattacks on foreign targets that threaten key U.S. assets such as power grids.[19] Authority for the U.S. military to launch cyberattacks outside of war zones would run counter to U.S. efforts to persuade foreign governments to curb digital attacks that emanate from their soil, including digital attacks on dissidents.

Moreover, U.S. government monitoring of digital communications, while necessary to prevent terrorist attacks and to investigate criminal activity, is sometimes at odds with the protection of online privacy, which is integral to the freedoms of expression and association. This very same monitoring, which is subject to due process and thus is consistent with privacy protections in the United States and other democratic countries, leads to abuse in repressive regimes, where due process is absent. Technology for online surveillance, such as spyware, is used by authoritarian governments to harass and prosecute dissidents.

The Obama administration reportedly plans to seek legislation to require all communications service providers to become technically capable of complying with a court order to wiretap communications. This requirement would create "backdoors" to intercept encrypted messages on BlackBerry, social-networking sites such as Facebook, and peer-to-peer services such as Skype.[20] National-security officials argue that these backdoors are essential for implementing existing wiretap laws and averting terrorist attacks, but the backdoors would then be open for abuse by repressive regimes, which monitor the communications of dissidents without the kind of independent judicial oversight available in the United States. The creation of backdoors for wiretapping will greatly increase the difficulty for communications-service providers to resist pressure from authoritarian governments to intercept private communications and, as the experience of BlackBerry maker RIM shows, such pressure is likely to grow.

Economic Interests

Economic interests pull the U.S. government in different directions on Internet-freedom issues. Some companies build their business by expanding access to information or providing digital-communications services that are reliable and secure. They have a strong interest in avoiding censorship and protecting the privacy of their users, although they usually give in to pressure from repressive governments to censor content or grant access to private data so that they can maintain their share of foreign markets. Google's decision in January 2010 to stop censoring web searches in China was among the very few exceptions to the common practice of U.S. companies giving in to such pressure.[21]

Other U.S. technology companies sell security software and services and seek to expand their overseas sales, even at the risk of seeing their products used to filter or monitor political content. Internet service providers in the Middle East and North Africa, for example, rely on commercial software from Websense and Secure Computing (which is now part of McAfee) to filter political content.[22] Boeing subsidiary Narus sold technology to the state-run company Telecom Egypt to intercept and inspect Internet and mobile-phone communications.[23]

The Obama administration has generally been passive in responding to pressure from repressive regimes on U.S. companies to assist with Internet censorship or surveillance. It has considered this pressure as a problem for U.S. companies to solve alone. In her January 2010 speech on Internet freedom, Secretary Clinton urged "U.S. media companies to take a proactive role in challenging foreign governments' demands for censorship and surveillance," but she offered no assistance from the U.S. government, beyond a meeting with the undersecretaries of state for economic affairs and for democracy and global affairs, who lead the State Department's Global Internet Freedom Task Force.[24] U.S. companies cannot be expected to stand up to pressure from foreign governments on their own.

The U.S. Trade Representative has, with rare exceptions, failed to challenge Internet censorship on trade grounds. While U.S. trade agreements have begun to make reference to Internet freedom—so that they may serve as incentives for other governments to respect online freedom of expression—the U.S. government only once has seriously challenged Internet censorship as a restriction on trade. This was the challenge to the Green Dam content-control software, which the Chinese government had planned to mandate for preinstallation on all computers in 2009 until a domestic and international uproar caused the plan to be scrapped.

The Obama administration as a whole is reticent about pressuring U.S. companies to avoid cooperation with repressive regimes in Internet censorship and infringements on the rights of Internet users. It has encouraged participation in the Global Network Initiative (GNI), a voluntary code of conduct with self-enforcement mechanisms for technology companies, but steered clear of any legal mandates on U.S. companies. It has neither supported the Global Online Freedom Act[25]—the only bill before Congress that would require U.S. technology companies to disclose their censorship activities and protect the private data of foreign users—nor proposed an alternative.

Reliance on the GNI to ensure corporate good behavior is likely to have a limited effect. Three years after its launch in 2008, GNI still had only three technology companies among its members—Google, Microsoft, and Yahoo. GNI has yet to bring in corporate members based outside the United States, newer companies focused on Web 2.0 services (such as Facebook and Twitter), or companies that provide content filters or spyware, which are likely to contribute to violations of Internet freedom when in the hands of authoritarian governments. Furthermore, a voluntary code of conduct and self-enforcement mechanisms cannot protect companies, even large multinationals, from pressure by authoritarian regimes to filter content or to facilitate surveillance of Internet users. For example, under its deal with Chinese search engine Baidu, Microsoft is expected to censor search results in China.[26]

European Policy

European government policy on Internet freedom was initially articulated in mid-2010 by then–French foreign minister Bernard Kouchner and then–Dutch foreign minister Maxime Verhagen. They began to convene multistakeholder meetings with allied governments, technology companies, and human-rights groups to build consensus around initiatives to:

- Establish an international mechanism to monitor the commitments that countries have made regarding online freedom of expression;
- Draft a corporate code of conduct to prevent Internet filters from ending up in the hands of repressive regimes or, alternatively, persuade European companies to join the Global Network Initiative;
- Defend the rights of bloggers and cyberdissidents who face censorship, persecution, or imprisonment; and
- Codify the universal character of the Internet into law.[27]

Through the multistakeholder meetings and changes in European leadership on Internet-freedom issues (primarily the resignation of Bernard Kouchner and the engagement of Swedish foreign minister Carl Bildt), European policy was refined, and significant initiatives were launched. The Netherlands and Sweden have begun to fund programs to support bloggers and cyberdissidents who come under threat. They have also pushed for greater European Union funding for Internet-freedom programs. The European Commission's EuropeAid, in its 2011–13 strategy for democracy and human-rights programs, has included support for greater access to the Internet and freedom of expression online.

To encourage technology companies to respect human rights, European policy makers increasingly seek to build on existing mechanisms, particularly on the GNI, rather than develop a new corporate code of conduct. Dutch foreign minister Uri Rosenthal has gone further and suggested legislation along the lines of the Global Online Freedom Act proposed in the U.S. Congress.

To strengthen international respect for free expression on the Internet, European policy makers now stress existing human-rights principles rather than propose new international laws. Rosenthal pointed out that Article 19 of the Universal Declaration of Human Rights applies to the Internet, while Bildt stressed, "The same rights that people have offline—freedom of expression, including the freedom to seek information, freedom of assembly and association, amongst others—must also be protected online."[28]

The Swedish government supported the work of UN Special Rapporteur for Freedom of Expression Frank La Rue, who presented his report on Internet freedom to the UN Human Rights Council in June 2011. The report called attention to key challenges for Internet freedom, such as digital attacks on human-rights organizations, intermediary liability for online content, criminal penalties for legitimate online expression, and

"just-in-time" blocking of online content during elections, periods of social unrest, or other critical moments. It stressed several key principles
for an open Internet, declaring that Article 19 applies to the Internet;
that censorship should never be delegated to a private entity; and that
any limitation on freedom of expression must be defined by law, be
necessary and proportional, and be applied by an independent body with
safeguards against abuse. La Rue's report put forward a series of recommendations, which the Swedish and other democratic governments
pledged to promote.

European policy is generally in sync with U.S. policy on Internet freedom. European leaders speak out in defense of bloggers and cyberdissidents; fund programs to expand online freedom of expression; work to
build recognition and respect for international human-rights principles
that apply to the Internet; and support a multistakeholder and self-regulat-
ing approach to Internet governance.

European policy on Internet freedom suffers from the same limitations
as U.S. policy. European governments promote broad principles of Internet freedom and support cyberdissidents but have yet to seriously challenge repressive regimes over specific restrictions on digital communications and media. They have generally neglected to support technology
companies in resisting pressure from repressive regimes to cooperate with
Internet censorship and surveillance.[29] They look to European companies
to adopt a voluntary code of conduct and have yet to enact legislation to
curb the use of European technology to commit human-rights violations.
A European version of the Global Online Freedom Act was introduced in
the European Parliament in 2008 with the support of four political-party
groups, ranging from the center-right European People's Party to the So-
cialists and the Greens, but was opposed by the EU's Commissioner for
Telecommunications and was never passed.

In April 2011, the European Parliament voted to introduce export controls on technologies for monitoring mobile-phone and Internet use, but
these export controls still require the European Council's approval. A voluntary code of conduct is unlikely to avert a future case—similar to the
Nokia-Siemens sale of a monitoring center to Iran Telecom—where European technology is used to commit human-rights abuses.[30]

Nokia-Siemens came under harsh criticism in the European Parliament
for its sale to Iran Telecom, and there were calls for a ban on exports of
surveillance technology to Iran and other repressive regimes, but no export controls were introduced. A year later, several other European companies were found to have sold spyware to repressive regimes. Britain's
Gamma International provided its product FinSpy to Egypt's security service, which used the product to monitor online activities of dissidents.
This spyware infects the computers of dissidents and allows the security
service to capture key strokes and intercept audio streams, even when
the dissidents are using encrypted email or voice communications such

as Skype. In addition, Italian company HackingTeam has sold software that bypasses Skype's encryption and captures audio streams from a computer's memory to security agencies in the Middle East and North Africa.

Future International Policy

The Obama administration has developed a sophisticated, multifaceted policy to protect and advance Internet freedom. This policy reflects a serious U.S. commitment to support online freedom of expression and to push back on the restrictions imposed by authoritarian regimes. It promotes respect for international human-rights principles of free expression, defends the open multistakeholder structure of global Internet governance, supports digital activists, and makes substantial funding available to develop and apply new technologies to curb Internet censorship and surveillance. The Dutch and Swedish governments have launched similar policy initiatives and led the debate in the European Union to step up efforts to promote Internet freedom.

Current international policy on Internet freedom, however, is insufficient to stem the rising tide of authoritarian censorship and control over digital media and communications around the world. The proliferation of these controls is outpacing the efforts of citizens and democratic governments to expand the space for free expression on the Internet. The United States and the EU can have greater impact in their efforts to push back against Internet-related repression by challenging restrictive Internet laws and practices abroad, standing up for technology companies that are pressured into assisting authoritarian regimes with Internet censorship or surveillance, and enacting legislation to curb sales of U.S. and European technology that is used to violate human rights.[31]

The United States and the European Union should conduct targeted diplomatic initiatives to challenge restrictive Internet laws and practices. These initiatives should seek to avert new constraints before they are introduced and to persuade foreign governments to reconsider existing controls. Senior U.S. and European officials should raise concerns about such restrictions with their foreign counterparts and express them publicly to the media. In addition, the U.S. and the EU should challenge restrictive Internet laws and practices as barriers to trade wherever they can make this case and, when possible, take their disputes against such laws and practices to the World Trade Organization.

U.S. and European companies that come under pressure to facilitate violations of human rights—for instance, to provide access to private data or communications of their users—should have recourse to diplomatic support from their government. They usually are unable to withstand such pressure on their own. The United States and the European Union should stand ready to intervene if asked and to push back on the government exerting this pressure.

The Global Network Initiative or other voluntary measures are clearly inadequate to curb the export of U.S. technologies that are used to commit human-rights abuses. Some form of legislation is needed. Legislation should, at a minimum, mandate transparency of actions by U.S. companies that contribute to Internet censorship or surveillance. U.S. companies should disclose details of the requests they receive from foreign governments to filter web content, to turn over personal data of users, or to allow communications interception, and should report to the State Department their sales of filtering or monitoring technology to countries that restrict the Internet. Alternatively, the U.S. Congress might consider carefully crafted export controls on specific technologies to Internet-restricting countries. These export controls would target specific technologies, such as content filters or spyware, that serve the primary purpose of limiting flows of online information or monitoring private digital communications and thus are likely to be used by repressive regimes to violate human rights.

The U.S. government needs to balance its promotion of Internet freedom with often competing political, economic, or security interests. While these competing interests tend to take precedence, the U.S. government should look to tilt the balance more toward Internet freedom. In addressing cybersecurity concerns, for example, the U.S. government should avoid measures that weaken the open and unitary structure of the global Internet or facilitate online surveillance by repressive regimes—for instance by giving repressive regimes back-door access to private digital communications. Security requirements on Internet-related services, such as those assisting in monitoring terrorist suspects, should be narrowly defined, commensurate with the security risk, applied transparently (with strong safeguards against abuse), and subject to independent oversight.

U.S. and European support for Internet freedom will have greater impact if it is part of a robust policy to promote human rights and democracy in general. It cannot be a substitute for democracy assistance for local activism, government reforms, and pluralist political processes. It is instead a component of larger efforts to protect fundamental rights, foster greater citizen participation, and make governments more accountable. Internet freedom is far more likely to flourish in countries where fundamental rights are respected and government is responsive to citizens. U.S. and European policy should therefore seek to promote Internet freedom as a complement to a broader agenda of supporting human rights and democracy.

Every aspect of U.S. policy on Internet freedom would be more effective if conducted in concert with democratic allies. Joint diplomatic initiatives would make greater progress in promoting respect for international principles of free expression, defending bloggers and cyberactivists who come under threat, reducing Internet censorship, and challenging restrictive Internet laws and practices. Coordination on funding would avoid duplication, allow U.S. and European governments to focus on their comparative advantages, and reinforce common efforts. Initiatives to curb exports of technology that is used to commit human-rights violations would have greater impact if transparency requirements and export controls were applied equally to companies in all democratic countries.

To advance Internet freedom in the face of growing restrictions around the world, the United States and the European Union need to do more. They cannot rely entirely on uncontroversial measures, such as advocating broad principles, criticizing flagrant abuses, and funding programs that promote Internet freedom. They need to take bolder actions, such as challenging foreign governments on their restrictive Internet laws and controlling exports of U.S. and European monitoring technology to repressive regimes. Such actions are vital to reversing the global trend toward suppression of Internet freedom.

NOTES

The author thanks Sarah Cook, Richard Fontaine, Robert Guerra, Sanja Kelly, Caroline Nellemann, and Christopher Walker for their comments on earlier drafts. Research was updated for the period up to 12 July 2011.

1. Bernard Kouchner, "The Battle for the Internet," *New York Times*, 13 May 2010.

2. Internet freedom is defined here as the freedoms of expression and association and the free flow of information on the Internet and in digital communications, including mobile telephones.

3. *Freedom on the Net 2011*, Freedom House, available at *www.freedomhouse.org/report/freedom-net/freedom-net-2011*.

4. Shahin Abbasov, "Azerbaijan: Baku Moving to Restrict Online Free Speech?" EurasiaNet.org, 25 May 2011, available at *www.eurasianet.org/print/63554*.

5. "Iran Vows to Unplug Internet and Create a Private Network," *Wall Street Journal Asia*, 30 May 2011.

6. Iran country report, *Freedom on the Net 2011*, 7, available at *www.freedomhouse.org/images/File/FotN/Iran2011.pdf*.

7. RIM refused to release details of its agreement with the UAE's Telecommunications Regulatory Authority, but the agreement was reportedly "compatible with the UAE's regulatory framework." Lance Whitney, "RIM Averts BlackBerry Ban in UAE," CNET News, 8 October 2010, available at *http://news.cnet.com/8301-1009_3-20019011-83.html*.

8. Report by Wang Chen, "Concerning the Development and Administration of Our Country's Internet," trans. Human Rights in China, *China Rights Forum 2010*, no. 2, 2010.

9. Hillary Rodham Clinton, speech, "Internet Rights and Wrongs: Choices and Challenges in a Networked World," George Washington University, 15 February 2011.

10. See comments by Michael H. Posner, assistant secretary of state for democracy, human rights, and labor, in "Conversations with America: The State Department's Internet Freedom Strategy," 18 February 2011, available at *www.state.gov/j/drl/rls/rm/2011/157089.htm*.

11. Clinton, "Internet Rights and Wrongs."

12. Letter from Assistant Secretary Michael Posner, 24 May 2011.

13. See Broadcasting Board of Governors, *Fiscal Year 2012 Budget Request*, 88–89, available at *http://media.voanews.com/documents/FY_2012_BBG_Congressioal_Budget_ Final_Web_Version2.pdf*. In addition to the funds for Internet-censorship circumvention embedded in BBG's budget, Congress was expected to appropriate an additional $10 million in 2011 to the BBG for anticensorship technologies.

14. Funding for Internet freedom–related work is included in several awards for media and civil society from the U.S. Agency for International Development.

15. Clinton, "Internet Rights and Wrongs."

16. Mary Beth Sheridan, "Congress Trims State's Internet Freedom Funds," *Washington Post*, 13 April 2011.

17. Haystack was fast-tracked for a license to export to Iran with U.S. State Department support and was praised by Secretary Clinton. See Evgeny Morozov, "The Great Internet Freedom Fraud," *Slate*, 16 September 2010.

18. Rami G. Khouri, "When Arabs Tweet," *New York Times*, 22 July 2010. A similar argument was presented in Sami Ben Gharbia's blog post, "The Internet Freedom Fallacy and the Arab Digital Activism," 17 September 2010.

19. Ellen Nakashima, "Pentagon's Cyber Command Seeks Authority to Expand Its Battlefield," *Washington Post*, 6 November 2010.

20. Charlie Savage, "U.S. Tries to Make It Easier to Wiretap the Internet," *New York Times*, 27 September 2010.

21. Domain-name registration companies GoDaddy and Network Solutions followed Google's lead soon after.

22. Helmi Noman, "Middle East Censors Use Western Technologies to Block Viruses and Free Speech," OpenNet Initiative, 27 July 2009, available at *http://opennet. net/blog/2009/07/middle-east-censors-use-western-technologies-block-viruses-and-free- speech*.

23. Timothy Karr, "One U.S. Corporation's Role in Egypt's Brutal Crackdown," *Huffington Post*, 28 January 2011, available at *www.huffingtonpost.com/timothy-karr/one-us- corporations-role-_b_815281.html*.

24. In her February 2011 speech, Clinton similarly offered no support for U.S. companies facing pressure from authoritarian regimes to collaborate in Internet censorship or surveillance. She merely praised the Global Network Initiative and called for "strong corporate partners."

25. While other bills would introduce or strengthen sanctions on specific countries, the Global Online Freedom Act, H.R.1389, would apply to all Internet-restricting countries. It would require U.S. companies to disclose the methods of filtering they use and the content they block at the request of repressive regimes, to host personal data on servers outside of Internet-restricting countries, and to refer requests for personal data from these regimes to the U.S. Justice Department.

26. David Barboza, "Microsoft to Partner with China's Leading Search Engine," *New York Times*, 4 July 2011.

27. "The Internet and Freedom of Expression," communiqué issued by the Ministry of Foreign and European Affairs, Paris, 8 July 2010, see *http://ambafrance-us.org/spip.php?article1729*.

28. Carl Bildt, "Freedom of Expression on the Internet Cross-Regional Statement," speech to the Human Rights Council, 10 June 2011, available at *www.sweden.gov.se/sb/d/14194/a/170566*.

29. In February 2008, the European Parliament passed a proposal to treat Internet censorship by other countries as a trade barrier (see *www.webpronews.com/eu-to-consider-Internet-censorship-a-trade-barrier-2008-02*), but this proposal was never adopted by the European Council. In early 2010, U.S. and EU officials pressed China to drop its plans to require local certification of encryption-related products, such as firewalls, secure routers, and smart cards, for government tenders. This certification requirement would have obliged U.S. and European companies to submit to Chinese government testing and perhaps to pass on software source codes and other confidential information. The companies seemed to object less to the risk that this testing would contribute to increased Chinese surveillance of Internet activity than to the possibility of losing out on Chinese government contracts or having their intellectual property handed over to Chinese competitors. See Kathrin Hille, "IT Groups Warn Chinese on Regulation," *Financial Times*, 21 February 2010.

30. A joint venture of the Finnish mobile-phone giant Nokia and the German electronics and electrical-engineering company Siemens in 2008 sold Iran Telecom a monitoring center that allowed the Iranian Revolutionary Guard Corps to tap mobile phones and monitor electronic data transmissions. Human-rights advocates and intelligence officials believe that this center's electronic surveillance system was used to target dissidents. See Christopher Rhoads and Loretta Chao, "Iran's Web Spying Aided by Western Technology," *Wall Street Journal*, 22 June 2009.

31. For a comprehensive and insightful analysis of U.S. policy on Internet freedom, see Richard Fontaine and Will Rogers, "Internet Freedom: A Foreign Policy Imperative in the Digital Age," Center for a New American Security, June 2011.

INDEX